NATIONAL DEFENS

T0119431

Equipping the 21st Century Marine Corps

Alternative Equipping Strategies for Task-Organized Units

Joslyn Fleming, Jonathan P. Wong, Matthew W. Lewis,
Duncan Long, Angelena Bohman, Kathryn Connor, Rick Eden,
Michael J. Gaines

Prepared for the United States Marine Corps

For more information on this publication, visit www.rand.org/t/RR2822

Library of Congress Cataloging-in-Publication Data is available for this publication.
ISBN: 978-1-9774-0213-4

Published by the RAND Corporation, Santa Monica, Calif.
© Copyright 2021 RAND Corporation
RAND® is a registered trademark.

Cover: U.S. Marine Corps photo by Cpl. Bethanie Ryan.

Support RAND
Make a tax-deductible charitable contribution at
www.rand.org/giving/contribute

www.rand.org

Preface

In response to global unrest, the Marine Corps has implemented proactive measures to respond to threats. These measures include deploying task-organized units, also referred to as provisional units, to respond to combatant commander demands. While in the past these demands were usually met by more traditional force packages such as Marine Expeditionary Units (MEUs), fiscal constraints such as the underfunding of amphibious shipping requirements have blunted the effectiveness of the Marine Corps' preferred method of projecting power through expeditionary forces.

The result is a growing demand for task-organized, provisional units. Like regular units, these units are manned, trained, and equipped to conduct a myriad of missions across the range of military operations. However, their temporary nature and provisional missions are at odds with the way that the Marine Corps normally deploys units, therefore placing a burden on the enterprise.

This report documents the extent of the equipping challenges of provisional units and identifies and tests recommendations to mitigate those challenges.

The research reported here was completed in November 2018 and underwent security review with the sponsor and the Defense Office of Prepublication and Security Review before public release. This research was sponsored by the United States Marine Corps Operations Analysis Directorate and conducted within the Acquisition and Technology Policy Center of the RAND National Security Research Division (NSRD), which operates the National Defense Research Institute

(NDRI), a federally funded research and development center sponsored by the Office of the Secretary of Defense, the Joint Staff, the Unified Combatant Commands, the Navy, the Marine Corps, the defense agencies, and the defense intelligence enterprise.

For more information on the RAND Acquisition and Technology Policy Center, see www.rand.org/nsrd/atp or contact the director (contact information is provided on the web page).

Contents

Figures

Tables

Summary

Introduction

In recent years, the United States Marine Corps (USMC) has used a variety of task-organized provisional units to meet increasing mission demands. These provisional units are defined as "service or combatant commander-directed temporary assembl[ies] of personnel and equipment organized for a limited time for accomplishment of . . . specific mission[s]."[1] Like regular units, these units are manned, trained, and equipped to conduct a myriad of missions across the range of military operations.[2] However, they are assembled in nontraditional ways from personnel and equipment sources across the Marine Corps. Once equipment is sourced, the materiel is usually left in place while personnel rotate.

While the Marine Corps has deployed these types of units for many years, especially for short duration missions such as humanitarian assistance or disaster recovery, since 2012 the Marine Corps has experienced challenges associated with provisional units becoming de

---

[1] The definition is taken from the "Glossary" in Headquarters Marine Corps, *Management of Property in the Possession of the Marine Corps*, Marine Corps Order 4400.201 Volume 3, Washington, D.C., June 13, 2016 (hereafter MCO 4400.201).

[2] These units include (but are not limited to) Special Purpose Marine Air Ground Task Force-Crisis Response-Central Command (SPMAGTF-CR-CC), Task Force-Southwest (TF-SW), SPMAGTF-Crisis Response-Africa, Marine Rotational Force Europe (MRF-E), Black Sea Rotational Force (BSRF), Marine Rotational Force-Darwin (MRF-D), and SPMAGTF-Southern Command.

facto standing units. Provisional units often operate in small, independent units, which differs from traditional Marine Corps formations. This creates an outsize demand for equipment, particularly specialized equipment, compared to a similarly sized regular unit. Some materiel requirements—for command and control equipment, for instance— do not "scale down" as personnel are reduced. A piece of gear that could support 500 Marines is still required if only 100 are deployed. Additionally, the temporary nature of these units removes them from the normal equipment accountability, maintenance, and modernization processes the Marine Corps uses.

These issues have challenged the service's ability to make equipment available for these units. Organizations across the Marine Corps have reported to Headquarters Marine Corps, Installations and Logistics (HQMC, I&L) a variety of these issues. For instance, satellite communications systems are in short supply but are essential to many deploying units. Sourcing this equipment for provisional missions has proved difficult. Assigning responsibility for equipment maintenance and modernization has also been a challenge.

Moreover, efforts to equip provisional units properly impacts regular units as well. Traditional units must furnish equipment to provisional units, which is reportable as a readiness metric and can render them unable to fulfill contingency mission obligations. Maintenance responsibilities are also unclear, resulting in regular units being held accountable for them. Although the Marine Corps has successfully employed provisional units for some time now, these challenges require systematic validation and possibly corrective actions to improve readiness, accountability, and other challenges.[3]

---

[3] It should be noted that there are many other closely associated issues with provisional units across doctrine, organization, training, materiel, leadership and education, personnel, and facilities (DOTMLPF). Many equipping issues are closely related to personnel and training issues. For instance, the high demand for communications and electronic equipment comes with an associated high demand for specialized military occupational specialties (MOS) that are also in short supply. Other associated issues are added training requirements and rapid obsolescence of technology. However, the focus of this study was on equipping challenges, and those aspects were outside the scope of the project design.

Key Findings

At the request of HQMC, I&L, the RAND National Defense Research Institute undertook a systematic examination of provisional unit equipping. We used quantitative analysis of equipping data and an extensive set of semistructured interviews to validate and further characterize the problems of which HQMC, I&L has been made aware of. Through this research, five problem areas became evident:

- There is a challenge of managing risk when trading off support for provisional units and other Global Force Management commitments.
- Provisional units have high demands for certain equipment classes.
- The clarity of Marine Corps policies on provisional units could be improved.
- Information technology (IT) systems are less responsive to provisional unit needs.
- Accurate utilization rates are lacking to inform equipping decisions.

The Risk Trade-Off Between Provisional Units and Readiness

One of the fundamental issues identified during the analysis was the tension between supporting provisional units while ensuring readiness to execute major combat. Provisional units serve an important purpose, and the Marine Corps is currently committed to these units. To outfit provisional units, however, the Marine Corps' readiness to respond to other issues suffers. Further, to support the National Defense Strategy (NDS), the Marine Corps has prioritized preserving readiness for major combat operations rather than risk it on support for currently employed provisional units. A primary example is the Marine Corps' decision to restrict provisional units' access to war reserve materiel as an equipping solution. While most of the stakeholders we talked to understand the importance of preserving war reserve materiel, they were also concerned that this reticence may put provisional units at risk.

Provisional Units Have High Demands for Certain Equipment

In aggregate, provisional unit equipping needs seem modest: They comprise 1 percent of the total Marine Corps–approved acquisition objective (AAO) by dollar value.[4] Moreover, most provisional unit equipment requirements can be sourced from across the Marine Corps using materiel excess to the USMC-wide planned AAO. We found, however, that equipping provisional units placed strain on the Marine Corps for some types of equipment, and that the demand for some items is high relative to the AAO for the Marine Expeditionary Force (MEF) sourcing the provisional unit.

The provisional unit equipment density lists (EDLs) have between them 154 unique items of equipment (Table of Authorized Materiel Control Numbers [TAMCNs]) that have an AAO or requirement greater than zero and that are readiness reportable.[5] Most of these TAMCNs do not seem to present a significant burden: Approximately 80 percent have an aggregate EDL requirement of less than 5 percent of the USMC-wide AAO. For just six specific pieces of equipment, the aggregated provisional unit equipment requirement is greater than 10 percent of the USMC AAO.

When comparing provisional unit requirements to the primary sourcing MEF, however, we see that a significantly greater percentage of TAMCNs fall in the 10 percent or greater range. For example, II MEF is the primary source for Special Purpose Marine Air Ground Task Force-Crisis Response-Africa, Black Sea Rotational Force, and Task Force-Southwest. The EDLs for those provisional units have between them 114 unique, readiness reportable TAMCNs. When the requirements for those three EDLs are combined for each TAMCN, the quantity is over 10 percent of II MEF's AAO for 22 items, or approximately

---

[4] Analysis in this section is based on data provided to RAND by HQMC. EDL requirements are from Headquarters Marine Corps, "Task-Organized Unit EDLs.xlsx," HQMC, Installations and Logistics, October 3, 2017. U.S. Marine Corps and Marine Expeditionary Force (MEF) requirements were also provided to RAND; see Headquarters Marine Corps, "Force 2025 AAO vs Inventory UIC Level-Data Model.xlsm," HQMC, Installations and Logistics, October 19, 2017.

[5] According to Headquarters Marine Corps, *Marine Corps Readiness Reportable Ground Equipment*, Marine Corps Bulletin 3000, Washington, D.C., March 21, 2017.

20 percent of the 114 items. EDL requirements sourced primarily by I MEF and III MEF similarly seem to represent a greater burden than USMC-wide numbers would suggest.

As might be expected, this dynamic extends to those ten TAMCNs the USMC currently identifies as High Demand/Low Density (HD/LD) across the service.[6] Eight of the ten are present on at least one provisional unit EDL. For one of the MEFs, the HD/LD TAMCNs required to fulfill their provisional unit EDL requirements make up at least 10 percent of the MEF's equipment on hand.[7]

Marine Corps Policies on Provisional Units Are Not as Clear as Possible

Marine Corps policies lack clear, consistent guidance regarding provisional unit equipping. Most policies are not supportive of the needs of these units. Where policies are written, they lack clarity and are not well understood by the force they intend to assist. Marine Corps logistics (supply, maintenance, distribution, etc.) policies that account for provisional units for the most part focus on the establishment of the unit (EDL creation and validation) and readiness reporting of equipment. The primary gap in policy is what is required for the sustainment of these units as they become enduring requirements outside of the standard Global Force Management (GFM) and AAO processes. Another challenge is that many of the systems referenced in policy are not able to support the requirements of provisional units, which results in a disconnect between how accountability is supposed to work according to policy and how it works in practice.

Information Systems Are Not Responsive to Provisional Unit Needs

To maintain equipment accountability, the Marine Corps uses Total Force Structure Management System (TFSMS) to track authorized allowances and Global Combat Support System-Marine Corps

---

[6] Document provided to RAND by email from Headquarters Marine Corps, "HD/LD Equipment List.pptx," HQMC, Installation and Logistics, February 7, 2018.

[7] Unless otherwise noted, we used "usable EAP," or usable enterprise asset posture, as equipment on hand.

(GCSS-MC) to track on-hand quantities. While these systems and their associated processes generally work for standard Marine Corps units, the workarounds and adapted processes to use these systems for provisional units are cumbersome and fail to meet their needs. As a result, most units are defaulting to the use of locally maintained spreadsheets to maintain visibility for on-hand equipment. This negates many of the advantages of enterprise IT systems like TFSMS and Global Combat Support System-Marine Corps (GCSS-MC) because they do not enable visibility, tracking, and analysis.

It Is Difficult to Assess the Extent of the Problem due to Data Opacity

Equipment needs of provisional units are not well understood. Because these units are relatively new and take on novel missions, estimates for equipment needs require significant subjective judgment. As the Marine Corps continues to employ provisional units, though, it will be difficult to refine estimates of equipment need without usage data. None of the units interviewed were able to provide accurate assessments of utilization rates. Understanding utilization rates is critical for commanders to make informed decisions about which pieces of equipment are required or not required for the mission.

Conclusions and Recommendations

In documenting the extent of provisional units' equipping challenges, the research team hypothesized that a small number of underlying issues were causing most of the overall challenges that were initially identified by HQMC, I&L. However, this was not the case. Equipping problems stemmed from a large number of small problems that collectively caused negative impacts. We realized that no single course of action would significantly improve provisional unit equipping. We therefore determined that a multipronged approach to solve the totality of the identified problems was the best course of action.

A hybrid approach taking the most popular aspects of the analyzed courses of action is recommended. It consists of

- centralized management functions at Marine Corps Logistics Command (MARCORLOGCOM) to manage excess equipment inventory and remove the burden of redistribution from the MEFs
- selective procurement of additional HD/LD items to alleviate demand
- continued use of equipment sets in forward-deployed locations— MEU Augmentation Pool-Kuwait (MAP-K), Marine Corps Prepositioning Program-Norway (MCPP-N), Maritime Prepositioning Ships (MPS)—while simultaneously making it easier to pull from these pillars of the AAO.

We used a systematic approach to identify and test recommendations to mitigate provisional equipment challenges. These recommendations are informed by three imperatives: balance provisional unit equipping with overall readiness; minimize disruption to current Marine Corps practices; and accommodate provisional unit equipping needs while keeping policies and practices flexible enough to accommodate future needs. They are mutually supportive but independent. These recommendations should be implemented in a sequenced and systematic manner.

The recommendations are to

- institute a multipronged approach using risk matrices to prioritize unit equipping in the Global Force Management process
- employ more empirical assessments of unit mission and requirements
- update Marine Corps policies that account for provisional and rotational unit equipping
- develop equipping strategies that emphasize forward positioning and local acquisition
- improve communication and visibility across IT systems
- eliminate the principal end item (PEI) rotation policy.

Acknowledgments

The authors wish to acknowledge the considerable assistance provided by our sponsor, Richard Hicks from Marine Corps Installation and Logistics, and our study monitors, Major Holly Zabinski and Rick Clinger of the Marine Corps Operations Analysis Directorate. They provided tremendous support throughout this study.

We are also grateful to the other members of the Study Advisory Committee and other stakeholders for their input and guidance. We wish to thank the many interviewees who made themselves available and provided us with their input. We would like to especially thank John Turner from Marine Forces Pacific, who went above and beyond supporting our team in our trip to Marine Forces Pacific and providing us with a wealth of data. Additionally, thanks to Lieutenant Colonel Matt Emborsky, Ground Equipment Staging Program Officer in Charge, for hosting our team in Darwin, Australia, during the middle of Cyclone Marcus.

We thank our RAND colleagues Chris Mouton and Mike Decker, and our dedicated reviewers, Randall Steeb and David Luckey, for providing helpful advice and feedback on this research, as well as Michael Shurkin and Joshua Klimas for their input to this project.

Abbreviations

AAO	approved acquisition objective
BSRF	Black Sea Rotational Force
CD&I	Combat Development and Integration
CMR	consolidated memorandum receipt
COA	course of action
CONUS	continental United States
CRSP	Combat Ready Storage Program
DRRS	Defense Readiness Reporting System
EAP	enterprise asset posture
EDL	equipment density list
FY	fiscal year
GCSS-MC	Global Combat Support System-Marine Corps
GFM	Global Force Management
HD/LD	High Demand/Low Density
HMMWV	High Mobility Multipurpose Wheeled Vehicle
HQMC	Headquarters Marine Corps
I&L	Installations and Logistics
IT	information technology
LOGCOM	Logistics Command
MAP-K	MEU Augmentation Pool-Kuwait
MARCENT	Marine Forces Central Command
MARCORLOGCOM	Marine Corps Logistics Command
MARFOR	Marine Forces

MCO	Marine Corps Order
MCPP-N	Marine Corps Prepositioning Program-Norway
MEF	Marine Expeditionary Force
MET	mission essential task
MEU	Marine Expeditionary Unit
MOS	Military Occupational Specialty
MPF	Maritime Prepositioning Force
MPS	Maritime Prepositioning Ships
MRAP	Mine-Resistant Ambush Protected
MRF	Marine Corps Rotational Force
MRF-D	Marine Corps Rotational Force-Darwin
MRF-E	Marine Corps Rotational Force-Europe
MTVR	Medium Tactical Vehicle Replacement
NIIN	National Item Identification Number
O&M	operations and maintenance
ORF	operational reserve float
PEI	principal end item
PP&O	Plans, Policies, and Operations
SME	subject matter expert
SPMAGTF	Special Purpose Marine Air Ground Task Force
TAMCN	Table of Authorized Materiel Control Number
T/E	Table of Equipment
TFSD	Total Force Structure Division
TFSMS	Total Force Structure Management System
TFSP	Total Force Structure Process
TF-SW	Task Force-Southwest
TLCM-OST	Total Life Cycle Management-Operational Support Tool
TO&E	Table of Organization and Equipment
UFC	Unified Facilities Criteria
UIC	unit identification code
USMC	United States Marine Corps
USTRANSCOM	United States Transportation Command

CHAPTER ONE

Introduction

Background

In response to global unrest, characterized by events and crises such as the 2012 embassy attack in Benghazi that have outsized strategic and geopolitical impacts, the United States Marine Corps (USMC) has implemented proactive measures to respond to these threats. This includes deploying task-organized units, also referred to as provisional units, to respond to combatant commander demands. While in the past these demands were usually met by more traditional force packages, such as Marine Expeditionary Units (MEUs), fiscal constraints such as the underfunding of amphibious shipping requirements, have reduced the availability of the Marine Corps' preferred method of projecting power through expeditionary forces.[1] Additionally, the Marine Corps has pivoted from fighting a counterinsurgency and irregular warfare focus to prioritizing a major conflict posture against a near peer competitor. This has resulted in a reexamination of how the Marine Corps is currently postured.

The trend has been a growing demand for task-organized, provisional units. A provisional unit is defined as "a service or combatant

---

[1]   The Commandant of the Marine Corps spoke of the shortfall in amphibious shipping and its impact on combatant commander requirements for power projection before the House Appropriations Subcommittee on Defense in March 2018. See United States House of Representatives, *Statement of General Robert B. Neller Commandant of the Marine Corps: Hearing Before the House Appropriations Subcommittee on Defense on the Posture of the United States Marine Corps*, Washington, D.C., March 7, 2018.

commander-directed temporary assembly of personnel and equipment organized for a limited period of time for accomplishment of a specific mission."[2] Like regular units, these units are manned, trained, and equipped to conduct a myriad of missions across the range of military operations. However, their temporary nature and provisional missions are at odds with the way that the Marine Corps normally deploys units, therefore placing a burden on the enterprise.

Currently the Marine Corps employs several provisional units to support Global Force Management (GFM) requirements. These units include Special Purpose Marine Air Ground Task Force-Crisis Response-Central Command (SPMAGTF-CR-CC), Task Force-Southwest (TF-SW), SPMAGTF-Crisis Response-Africa (SPMAGTF-CR-AF), Marine Rotational Force-Europe (MRF-E), Black Sea Rotational Force (BSRF), Marine Rotational Force-Darwin (MRF-D), and SPMAGTF-Southern Command (SPMAGTF-SC).

Since provisional units are temporary organizations, there is little infrastructure and a lack of specific policy to validate and manage resources being used by these units. For example, provisional units do not have standing Tables of Organization and Equipment (TO&E), which list the authorized billets and equipment a unit possesses based on its mission. Instead, these units use equipment density lists (EDLs) and manning documents, which are determined at the commander's discretion, are not well tracked by information systems, and are rarely updated.[3] These documents are also not directly tied to the Marine Corps' force development process.

In addition, these units have become an enduring requirement over the last few decades of conflict and are projected to continue through fiscal year (FY) 2023, if not longer.[4] The permanent nature of these units, which were intended to be temporary, forces the Marine

---

[2]   Marine Corps Order (MCO) 4400.201, Management of Property in the Possession of the Marine Corps: Glossary, June 13, 2016.

[3]   This analysis is from interviews collected.

[4]   Summarized from personal email provided to RAND by Headquarters Marine Corps, "AAR- IPR #2 on RAND Analysis of Provisional Unit Equipment Management," Headquarters Marine Corps, Installation and Logistics (HQMC, I&L), May 9, 2018.

Corps to make risk calculations when it determines which units to prioritize when it comes to equipment and personnel. The requirements of these units are not specifically accounted for in determining approved acquisition objectives (AAOs). This results in a mismatch between the way the Marine Corps is currently employing the force and how it is equipping it.

Equipping Provisional Units

When a provisional unit rotates into theater, it follows a sequential equipping process. That process is outlined in MCO 4400.201 Volume 3. When interviewing personnel assigned to provisional units, we determined that the process adheres fairly well with only some slight variations. How the sequential process currently operates in practice is outlined below.

A provisional unit deploys, usually with personal gear.[5] The unit falls in on gear already in theater and conducts an equipment review no later than ninety days after its relief in place/transfer of authority (RIP/TOA). If the unit determines that changes to the EDL are required, the unit follows the review, validation, and approval process as outlined by the MCO.

The unit sends a request to its respective regionally aligned Marine Forces (MARFOR) Command for the equipment it requires. For example, SPMAGTF-CR-CC sends its request through U.S. Marine Forces Central Command (MARCENT) for any equipment it needs that is not currently on hand. The MARFOR first looks to the respective Marine Expeditionary Force (MEF), which provided forces for the unit, to lend the equipment needed. In the example of SPMAGTF-CR-CC, MARCENT would task I MEF to fill the requirement. If the MEF that originally sourced personnel and equipment cannot source, then the MARFORs and HQMC, I&L work together to find a global sourcing solution. Such solutions, according to the MCO, are

---

[5] According to our interviews, units usually deploy with personal weapons and optics, although guidance varies by command as to what is taken from home station.

1. Marine Corps Logistics Command (MARCORLOGCOM)
2. One of the other two MEFs
3. Preposition stock around the globe for use in wartime (e.g., MEU Augmentation Program-Kuwait [MAP-K], Marine Corps Pre-Positioning Program-Norway [MCPP-N]).

If nonexcess cross-leveling cannot occur, HQMC, Plans, Policies, and Operations (HQMC, PP&O) evaluates risk, and the Deputy Commandant, Combat Development and Integration (DC, CD&I) evaluates any AAO changes.

Once sourced, the requested equipment is then transported to the provisional unit. If coming from the continental United States (CONUS) it can take anywhere between two to seven months to arrive.[6] If coming from a forward-deployed location such as MAP-K, it can take as little as one to two weeks but can also be extended by months depending on which countries the equipment is moving through and the time it takes to process through customs and other foreign government restrictions.[7] The provisional unit then rotates the equipment to be replaced back to the MEF that sourced it originally for maintenance and modernization. When a new provisional unit is created, instead of falling in on equipment, the unit will be initially sourced by the same options listed above.

Figure 1.1 provides a geographical understanding of where various provisional units are located and from where their potential sourcing can come from. Highlighted specifically are the sourcing options, outlined above, for SPMAGTF-CR-CC located in Kuwait.

Sourcing personnel and equipment for provisional units has proved to be a challenge according to experts interviewed. Equipment is obtained from across the Marine Corps, not always in a systematic manner, which impacts transportation costs. Additionally, this reshuffling of equipment is not well documented, leading to gaps in AAO development and the modernization of Marine Corps equipment.

---

6   This information was provided during subject matter expert (SME) interviews.

7   This information was provided during SME interviews.

Figure 1.1
Special Purpose Marine Air Ground Task Force-Crisis Response-Central Command Sourcing Options

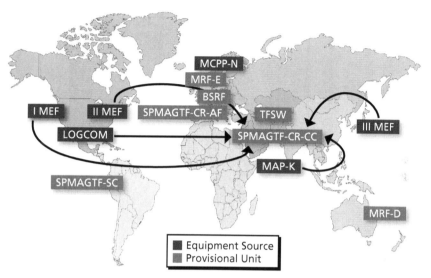

SOURCE: RAND analysis derived from Headquarters Marine Corps, *Management of Property in the Possession of the Marine Corps*, Marine Corps Order 4400.201 Volume 3, Washington, D.C., June 13, 2016, and subject matter expert interviews.

Once equipment is sourced, it is often left in theater for extended periods of time without required maintenance or modernization. Policies have been adopted in an ad hoc manner to address some of these concerns, such as the recent principal end item (PEI) rotation policy that requires 20 percent of a unit's equipment in theater to be sent back to the United States for upgrade during each rotation.

Even with recent policy changes, many challenges remain. The EDLs used to track provisional unit equipment are not always accurate, lack enterprise-wide visibility, and do not always align well with mission requirements. Home unit readiness is likely impacted but difficult to assess with current gaps in IT systems. Additionally, transporting equipment into and out of theater is very costly and time-consuming, which means either the sourcing unit or the deployed unit is without its needed equipment for an extended period of time.

To confirm and address all of these challenges, HQMC, I&L asked RAND to analyze how the Marine Corps currently equips the force and determine how best to adapt its equipping policies and processes to better match how forces are being employed into the twenty-first century and ultimately increase readiness across the force.

Research Approach

RAND's research approach consisted of two integrated tasks:

1. Document current provisional unit equipping practices.
2. Identify and assess the impact of alternative equipping processes on Marine Corps units and equipment.[8]

To understand the current equipping practices of the Marine Corps, we began with a document review of current Marine Corps policies, orders, and doctrine related to provisional units and equipping. We then conducted quantitative analysis of provisional unit EDLs. We compared EDL data to other standard unit equipment counts to understand key differences and ultimately determine what impact provisional unit equipping has on home unit equipment availability. Specifically, provisional units' use of High Demand/Low Density (HD/LD) equipment was analyzed.

This analysis was complemented by semistructured interviews with a range of provisional unit equipping stakeholders. These interviews helped us to both understand current equipping processes and impacts of potential alternatives and mitigation strategies. Overarching themes were extracted from the interviews using qualitative coding methodology, which is further described in Appendix A. The stakeholder organizations interviewed are listed in Table 1.1.[9]

---

[8] The study specifically excluded aviation equipment from the analysis per the sponsor's request.

[9] RAND's Human Subjects Protection Committee (HSPC) deemed this effort was not human subjects research and thus not subject to HSPC oversight. However, we still acted to

Table 1.1
Stakeholder Organizations Interviewed

Provisional Units	Headquarters Marine Corps	Operating Force	Supporting Establishment
Special Purpose Marine Air Ground Task Force-Crisis Response-Africa (SPMAGTF-CR-AF) Special Purpose Marine Air Ground Task Force-Crisis Response-Central Command (SPMAGTF-CR-CC) Marine Rotational Force-Darwin (MRF-D) Marine Rotational Force-Europe (MRF-E)	HQMC, I&L HQMC, PP&O Combat Development and Integration (HQMC, CD&I) Total Force Structure Division (HQMC, TFSD)	I MEF III MEF 7th Marines Marine Forces Pacific Command (MARFORPAC) 4th Tank Battalion (Reserve Unit) Combat Logistics Regiment-1 MARCENT Marine Forces Command (MARFORCOM)	MARCORLOGCOM Exercise Support Division (ESD)

SOURCE: RAND.

NOTE: Interviewees were determined in coordination with HQMC, I&L and the Marine Corps Operations Analysis Directorate.

We then assessed the challenges identified during the documentation phase of the project and developed solution elements based on previous RAND research, inferences from our own analyses, and inputs from the interviewees. These solution elements were then grouped into distinguishable courses of action that were further evaluated for costs and benefits. Cost data were used to assess personnel, maintenance, and transportation costs of the proposed courses of action.

Organization of Report

The remainder of this report presents our research. In Chapter Two, we further examine the equipping challenges experienced by both pro-

maintain confidentiality to protect the identities of respondents, so interview quotes will not be identified by person or organization.

visional units and those units supporting them. Chapter Three out-lines how we created solution elements to address the challenges identified. These solution elements are then combined into courses of action. Those courses of action (policy enforcement, centralized management, permanent structure, and tailored equipping) are evaluated in Chapter Four, and Chapter Five presents final recommendations. Appendix A contains the protocols used during interviews. Appendix B details our qualitative methodology. Appendix C provides a rough order of magnitude cost analysis for each course of action that was considered.

CHAPTER TWO

The Challenges Created by Equipping Provisional Units

The current security environment requires a Marine Corps that is forward deployed and able to respond to crises in an expeditious manner. To meet these requirements the Marine Corps is postured to be responsive and scalable by employing multiple provisional units to meet combatant commander demands across the globe. These units are considered provisional because, with some exceptions, they are performing provisional missions and have provisional unit identification codes. They lack standing TO&Es and instead use manning documents and EDLs. The way these units are sourced and deployed is different from the normal GFM processes that have been established in Marine Corps doctrine and policies; this puts increased stress on the force.[1]

This chapter analyzes the current challenges facing provisional units as relates to their equipping. It will first discuss how challenges were identified and then describe them. Those challenges are

- tension between supporting provisional units at the expense of GFM requirements and the associated risks
- changes in concept of employment and its impact on AAOs
- lack of inclusion of provisional units in Marine Corps policies

---

[1] Summarized from United States Marine Corps Operations Analysis Directorate, "Performance Work Statement for the Analysis of Provisional Unit Equipment Management," HQMC, Combat Development and Integration, June 23, 2017.

- information technology (IT) systems that are not responsive to provisional unit needs
- lack of accurate utilization rates to inform equipping decisions.

How Challenges Were Identified

Our analytical framework for identifying challenges was to take the issues identified at the beginning of the project and validate those assumptions via a three-part process. During this process additional challenges were also identified and explored. From there we identified solution elements, which we then compiled into comprehensive courses of action for evaluation. The three-part process for identifying challenges included using logistics data to quantitatively analyze provisional unit EDLs and their impacts, interviewing SMEs, and conducting a review of current Marine Corps policies and orders.

We used data provided by I&L to analyze EDL equipment and its on-hand impact.[2] This included seven EDLs for six provisional units[3] (MRF-D, MRF-E/BSRF, SPMAGTF-CR-AF, SPMAGTF-CR-CC, Task Force 9.7, and TF-SW). These documents were current circa 2017 and were the most current versions available when the study was conducted. We also used I&L data on USMC-wide requirements and on-hand quantities. These data helped us to place provisional unit equipment demand in context and assess the impact that provisional unit equipping was having on equipment availability at home stations. It also led to insights about the impact of HD/LD availability, which will be discussed later in this chapter.

Simultaneously, we conducted a series of semistructured interviews with SMEs to identify challenges facing the force. We spoke to both the provisional units themselves and organizations that provided support for these units as well as headquarters elements responsible for

---

[2] HQMC, "Task-Organized Unit EDLs.xlsx," October 3, 2017.

[3] One unit, SPMAGTF-CR-AF, had two separate EDLs for the Command Element and the Logistics Combat Element, whereas the other units combined all subordinate units on one EDL.

writing policies and overseeing the equipping process. These interviews were then coded using Dedoose, a collaborative mixed-methods software that assists in data analysis. A more in-depth discussion of the interviews and the use of Dedoose is described in Appendix A. These interviews identified additional challenges and helped define solution elements outlined in Chapter Three.

Finally, we did a comprehensive document review of Marine Corps policies, orders, and directives pertaining to provisional units and equipping. This led to insights into how policies do not account for provisional units and led to recommendations about how these units can best be incorporated into existing Marine Corps policy and foundational documents.

Risk to Major Combat Readiness by Supporting Provisional Units

One of the fundamental issues identified during the analysis was the tension between supporting provisional units and the risk it poses to Marine Corps readiness to respond to major combat. To support the National Defense Strategy (NDS), the Marine Corps has prioritized preserving major combat operations readiness rather than risk it on support for currently employed provisional units.[4] A primary example is the Marine Corps' decision to preserve the war reserve pillar of the AAO, making it difficult for provisional units to draw equipment from prepositioned stocks. While stakeholders we interviewed understand the importance of preserving war stocks, there was concern that the "break glass in case of war" approach currently employed may put those currently deployed unnecessarily at risk. As one Marine put it,

---

[4] This analysis is summarized from personal email provided to RAND by Headquarters Marine Corps, "AAR- IPR #2 on RAND Analysis of Provisional Unit Equipment Management," HQMC, I&L, May 9, 2018.

Marines are going into harm's way while there are war reserves right there that we aren't allowed to use.[5]

The burden is placed on the supporting MARFORs and MEFs to identify solutions to shortfalls when new requirements for provisional unit equipment or modernization to current inventory is identified. This requires the MARFORs and MEFs to source equipment from home stations to prioritize these requirements. While this may be the appropriate risk calculation, the decisionmaking is decentralized, nonstandardized, and not always well captured at higher echelons of command.[6] Therefore, problems are not always identified early in the sourcing process. Lack of visibility by the enterprise as a whole also inhibits other units from learning from the lessons of others.

In some cases, the MARFORs and MEFs do not have the ability to source requirements, which adds further delay to the process. SMEs indicate that under the current process, it sometimes takes from two to seven months to source replacement equipment from their home station, whereas when units are allowed to use forward-prepositioned equipment, such as MAP-K and MCPP-N, the time line is shortened significantly to a few weeks at most.[7]

This prioritization of equipment is somewhat at odds with the way the Marine Corps currently employs its forces and how it plans to deploy them in the future. Marine Corps planners anticipate that the current rotational force employment model will endure through FY 2023 and beyond.[8] Marine Corps strategic documents also emphasize that the base unit of deployment will continue to get smaller and units will be more distributed on the battlefield.[9] We assess this future

---

[5]   We have included selected quotes from interview participants to better illustrate the findings of our study. To protect confidentiality and for the safety of participants, we do not identify or cross-reference the source for quotes.

[6]   This information was provided during SME interviews.

[7]   This information was provided during SME interviews.

[8]   This information was provided during SME interviews.

[9]   Headquarters Marine Corps, *Marine Corps Operating Concept*, Washington, D.C., June 2016.

concept will stress Marine Corps logistics and equipping practices in much the same way that provisional units are currently taxing the force.

Changes in Concept of Employment and Impact on AAOs

We assessed that one of the primary drivers of the provisional unit equipping problem is that provisional units are not factored into determining an item's AAO. The Marine Corps establishes acquisition objectives to equip permanent units and obligations (like war reserves). Provisional units are temporary and are thus not factored into the acquisition objective. Although the AAO "must be regularly reviewed and revalidated to ensure the AAO continues to reflect the concept of employment given the projected force structure and the changes in capability,"[10] there is no guidance to account for units that do not operate under a standard unit identification code (UIC). Similarly, when a unit with a standard UIC is assigned to a provisional unit, such as an infantry battalion, the tasks assigned differ from the unit's core tasks, necessitating different equipment. There is no systematic way to capture this change in utilization that is then communicated back to the Marine Corps supporting establishment for review.

Current Provisional Unit Employment

The current process of determining AAOs based on permanent unit utilization is at odds with the Marine Corps' recent trend of disaggregating and deploying smaller units. Recent employment concepts have routinely deployed company-size elements or smaller. Often, these units require enablers and other supporting assets that are normally held at higher echelons. To meet mission requirements, then, these smaller units require quantities of equipment disproportionate to their size.

Additionally, as units continue to rotate into geographic combatant commands, commanders have allowed their units to be used

---

[10] Headquarters Marine Corps, *Total Force Structure Process*, Marine Corps Order 5311.1E, Washington, D.C., November 18, 2015 (hereafter MCO 5311.1E).

in ways beyond the scope of their intended mission essential tasks (METs). As one interviewee related,

> *The combatant commander likes having forces, but not a lot of contingencies were actually happening. To justify the forces, they allowed mission creep to happen without going through the right process [GFM process]. Loosely defined requirements. We were finding missions for ourselves.*

Combatant commanders have a high demand for forces within their area of operations that are not always able to be met. Therefore, SMEs articulated that provisional units operating in those theaters can be tasked to fulfill requirements outside of their intended scope. This results in units requesting additional equipment to meet mission requirements they were not initially intended to fulfill. For instance, in Central Command, small units are being deployed in a highly distributed manner to conduct missions they were not initially designed to conduct, which exacerbates the demand for equipment. This is felt most profoundly on HD/LD assets such as strategic communication assets, intelligence assets, and centralized maintenance equipment.[11] The combination of increasing amounts of small units operating in a decentralized manner and performing missions beyond the scope of their assigned METs, all outside the designed equipment AAOs, results in a disparity between how the Marine Corps operates and how it equips.

High Demand/Low Density Appetite

Does this disparity result in equipment shortages and management problems for provisional units or the Marine Corps generally? It is conceivable that equipment surpluses and non-AAO assets (e.g., equipment acquired using rapid acquisition processes or funds from other overseas contingency operations sources) may result in enough equipment to mitigate the problem.

---

[11] This information was provided during SME interviews.

To determine the impact of resourcing provisional units on the Marine Corps enterprise, we analyzed recent provisional unit deployments and found that provisional equipment demands affect some types of equipment more deeply than others. In aggregate, provisional unit equipping needs comprise 1 percent of the total Marine Corps AAO by dollar value. Provisional unit demand for equipment does not exceed the full AAO for any Table of Authorized Materiel Control Numbers (TAMCNs) that have a reported AAO greater than zero. However, while it appears that most provisional unit EDL requirements can be sourced from across the Marine Corps without significant strain, the demand for some TAMCNs is high relative to the AAO for the MEF sourcing the provisional unit.

Management issues related to provisional unit equipping are also evident. A concern that many interviewees stated was that although the overall Marine Corps AAO can support the demand from the provisional units, the ability to pull from AAOs outside of the operational force was difficult. Equipment accounted for under the war reserves pillar of the AAO, such as items in MCPP-N or onboard Maritime Prepositioning Force (MPF) ships, is not being tasked to support provisional unit requirements. The Marine Corps is hesitant to tap into these stocks for rotational unit requirements because it risks those stocks not being available to support potential major combat operations. While this is a risk, some interviewees felt that it unnecessarily put Marines at risk who were currently operating at the "tip of the spear."

Most notably, a concern cited by interviewees about provisional unit equipping is that these units exacerbate issues the force faces with HD/LD items, adding another source of demand to pieces of equipment that are already too scarce to meet current requirements. Our review of USMC on-hand data indicates that provisional unit HD/LD requirements are a minor share of total requirements and seem to us to be a relatively minor contributor to overall USMC supply constraints. Provisional units *do*, however, have HD/LD TAMCNs on their EDLs and consequently contribute to the already-identified shortage of supply for some of these items. This impact could be acute if provisional units are sourced only from a single MEF.

We used data provided by I&L to conduct our analysis on provisional unit EDL equipment requirements. I&L furnished a spreadsheet, current as of approximately September 2017, that showed the EDLs for the six provisional units show in Table 2.1.[12] This source provided equipment on hand and equipment required for some units; for others, it showed only an unspecified quantity. We treated unspecified quantities as a stated requirement and grouped them with requirement values on the other EDLs.

I&L also provided USMC-wide equipment on hand and equipment required at the unit level.[13] The on-hand data were current as of October 2017.[14] The requirement data we applied were for FY 2018.[15]

We initially began our analysis using a 2016 USMC HD/LD list with 43 total TAMCNs. We were later supplied with a list current as of January 2016, with just ten items, five of which were on the earlier list.[16] Those ten items, shown in Table 2.2, are the basis for the following analysis. The list is heavily weighted toward communications and electronics equipment, and only half the items are Marine Corps Automated Readiness Evaluation System (MARES) reportable items for unit supply readiness purposes. It should be noted that all elements underlying the analysis—EDL demand, USMC-wide demand, HD/LD supply, and what TAMCNs are considered HD/LD—are subject to change over time.

By nature, HD/LD items are a small share of overall equipment both on the EDLs and in the USMC at large. The EDLs collectively have 591 unique TAMCNs, and eight of the ten HD/LD TAMCNs appear on at least one EDL. In total, the EDLs require about $8 mil-

---

[12] HQMC, "Task-Organized Unit EDLs.xlsx," October 3, 2017.

[13] HQMC, "Force 2025 AAO vs Inventory UIC Level-Data Model.xlsm," October 19, 2017. Unless otherwise noted, all descriptive information about TAMCNs is derived from this source.

[14] Unless otherwise noted, we used "usable EAP," or usable enterprise asset posture, as equipment on hand.

[15] Unless otherwise noted, we used "total AAO," or total approved acquisition objective, as the required quantity.

[16] HQMC, "HD/LD Equipment List.pptx," February 7, 2018.

Table 2.1
Provisional Unit EDLs Included in the Analysis

Provisional Unit Name	Abbreviation	Principal Sourcing MEF[1]
Marine Rotational Force-Darwin	MRF-D	III MEF
SPMAGTF-Crisis Response-Africa	SPMAGTF-CR-AF	II MEF
Black Sea Rotational Force Marine Rotational Force-Europe	BSRF MRF-E	II MEF
SPMAGTF-Crisis Response-Central Command	SPMAGTF-CR-CC	I MEF
Task Force 9.7	Task Force 9.7	I MEF
Task Force-Southwest	Task Force Southwest	II MEF

SOURCE: RAND.

NOTE: The Task Force 9.7 mission was discontinued during the study period. It is included in the analysis. The analysis does not include SPMAGTF-Southern Command, which was not in I&L's initial compilation of EDLs.

[1] Brian Turner et al., *Program Objective Memorandum-19 Front End Assessment; Aligning Ground Equipment Inventories to Operational Requirements*, Washington, D.C., September 30, 2016, p. 35 (supplemented by RAND analysis drawing on unclassified press reports).

lion in HD/LD TAMCNs, about 0.2 percent of their total $0.54 billion requirement. That $0.54 billion is in turn dwarfed by the USMC's total AAO of approximately $33.7 billion.

The HD/LD TAMCNs on the EDLs appear somewhat more consequential when considered as quantities of discrete items. As Table 2.3 shows, of the eight HD/LD items present on at least one EDL, the average EDL requirement is 7 percent of USMC-wide usable enterprise asset posture (EAP).[17] The EDLs thus place a slightly higher demand on HD/LD than on the extant equipment pool at large: The average for all items on the EDL, for those items with usable EAP greater than zero, is 5 percent of USMC-wide usable EAP.

[17] The AAO for these items is equal to or higher than the usable EAP. The EDLs average 5 percent of the USMC-wide AAO for the eight HD/LD items.

Table 2.2
HD/LD TAMCNs

TAMCN	COLLOQUIAL NAME	COMMODITY GROUP	REPORTABLE CATEGORY	FUNCTIONAL AREA (reportable items only)	Average Unit Price[1]
A0176	LAN EXTENSION MODULE	Comm Elec	PEI	Data Comm	$28,176
A0278	WIRELESS POINT TO POINT LINK (WPPL) D	Comm Elec	PEI	Radios	$100,000
A0304	INFORMATION ASSURANCE MODULE (IAM) DDS-M	Comm Elec	PEI	Data Comm	$50,000
A0364	ECCS RRK	Comm Elec	MEE	Sat Comm	$236,575
A2634	TCAC RAWS V4.6	Comm Elec	MEE	Surv Intel	$73,790
A7597	TEST SYSTEM, VIPERT EO	Comm Elec	Non Rpt		$915,679
E0117	JOINT TERMINAL ATTACK CONTROLLER LASER TARGET DESIGNATOR (JTACLTD)	Ordnance	Non Rpt		$52,875
E2658	TOOL KIT, IM, LAV-25, 3D ECH	Ordnance	Non Rpt		$26,031
E2659	TOOL KIT, IM, LAV-25, 4TH ECH	Ordnance	Non Rpt		$35,721
E3163	TOOL SET, FM, 3D/4TH ECH, F/AAV	Ordnance	Non Rpt		$29,000

SOURCE: HQMC, I&L.

NOTE: PEI: principal end item; MEE: mission essential equipment, a subset of PEI.

[1] We used the I&L data to derive an average standard unit price for each item. Each TAMCN may have more than one associated National Item Identification Number (NIIN), and each NIIN may have a different standard unit price. In order to conduct consistent analysis at the TAMCN level, we developed a TAMCN standard unit price based on the average of associated NIINs, weighted by the quantity of those NIINs currently on hand.

Table 2.3
HD/LD Equipment on EDLs Compared to Entire USMC

TAMCN	COLLOQUIAL NAME	Total EDL Quantity	EDL as % of USMC-wide Usable EAP Qty
A0176	LAN EXTENSION MODULE	87	5%
A0278	WIRELESS POINT TO POINT LINK (WPPL) D	18	7%
A0304	INFORMATION ASSURANCE MODULE (IAM) DDS-M	8	5%
A0364	ECCS RRK	2	4%
A2634	TCAC RAWS V4.6	3	13%
A7597	TEST SYSTEM, VIPERT EO	2	4%
E0117	JOINT TERMINAL ATTACK CONTROLLER LASER TARGET DESIGNATOR (JTACLTD)	12	16%
E2658	TOOL KIT, IM, LAV-25, 3D ECH	1	5%
Average, Items on EDL			7%

SOURCE: RAND analysis.

Another way to place the EDL HD/LD impact in context is to assess it in terms of the demand each EDL places on its principal sourcing MEF. Table 2.4 summarizes the EDL HD/LD burden for each MEF. The percentage for each MEF compares the total quantity from EDLs for which it is the primary source (see Table 2.1) with only that MEF's usable EAP.[18] While noting that EDL requirements can be sourced from across the USMC, this assessment shows that each HD/LD TAMCN places a burden of at least 10 percent of usable EAP on at least one MEF.

The HD/LD demand from EDLs is, again, comparatively minor in terms of overall value and not significantly different from the USMC-wide impact of all EDL items. It is clear, however, that EDLs

---

[18] MEF AAOs are equal to or higher than MEF usable EAP in these cases.

Table 2.4
EDL HD/LD Impact on Sourcing MEFs

TAMCN	COLLOQUIAL NAME	I MEF EDL Qty as % I MEF Usable EAP	II MEF EDL Qty as % II MEF Usable EAP	III MEF EDL Qty as % III MEF Usable EAP
A0176	LAN EXTENSION MODULE	10%	4%	2%
A0278	WIRELESS POINT TO POINT LINK (WPPL) D	27%	6%	0%
A0304	INFORMATION ASSURANCE MODULE (IAM) DDS-M	5%	17%	0%
A0364	ECCS RRK	0%	22%	0%
A2634	TCAC RAWS V4.6	17%	9%	0%
A7597	TEST SYSTEM, VIPERT EO	33%	0%	0%
E0117	JOINT TERMINAL ATTACK CONTROLLER LASER TARGET DESIGNATOR (JTACLTD)	67%	24%	0%
E2658	TOOL KIT, IM, LAV-25, 3D ECH	0%	25%	0%

SOURCE: RAND analysis.

add to demand for items already identified as scarce and that demand can be significant relative to MEF supply.

EDL Impact on Supply Readiness

Provisional unit EDL impact on HD/LD TAMCNs is echoed in the apparent EDL impact on USMC supply readiness at large. In aggregate, the EDL requirement is only a small share of the USMC's overall requirement. However, the impact is significantly greater for discrete TAMCNs and when compared to MEF-level supply.

Figure 2.1 illustrates this impact. The bar at far left shows the 154 TAMCNs that have an AAO and that are readiness reportable. The colors categorize the EDL TAMCNs, grouping them by the EDL requirement's share of the total USMC AAO. About 80 percent of the 154 unique EDL TAMCNs (the dark green and light green bands combined) have an EDL requirement of less than 5 percent of the USMC-wide AAO. For just six specific pieces of equipment, the aggregated

Figure 2.1
Quantity of Each EDL TAMCN Requirement as Percentage of AAO

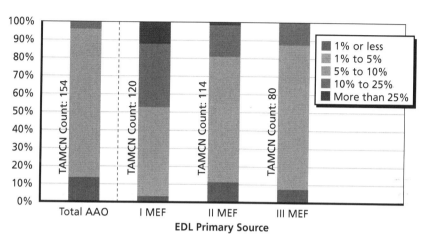

SOURCE: RAND analysis.

provisional unit equipment requirement is greater than 10 percent of the USMC AAO.

When comparing provisional unit requirements to the primary sourcing MEF, however, we see that a significantly greater percentage of TAMCNs fall in the 10 percent or greater range. A plurality of these TAMCNs are Communications and Electronics. As an example, II MEF is the primary source for SPMAGTF-CR-AF, BSRF, and TF-SW. The EDLs for those provisional units have between them 114 unique, readiness reportable TAMCNs. When the requirements for those three EDLs are combined for each TAMCN, the quantity is over 10 percent of II MEF's AAO (the orange and red bands combined) for about 20 percent of 114 items.

We extended this analysis by using the available unclassified equipment on-hand data to approximate the supply (or S-rating) impact on the MEFs sourcing the EDLs. The S-rating is calculated by dividing usable EAP by the AAO quantity for reportable items. A figure greater than or equal to 90 percent is a rating of S-1. We assumed that the EDLs were sourced entirely by their primary MEF, and further that the MEFs' requirements were not backfilled. We then set out to estimate the number of EDL TAMCNs that would be S-1 in the MEFs but for the need to source the EDLs.

For EDLs sourced by I MEF and II MEF, the equipment on hand at the provisional unit is no longer associated with the MEF but with one of the MARFORs—Marine Forces Europe/Africa (MARFOREUR/AF), MARCENT—depending on the provisional unit. To assess the impact, we added provisional unit usable EAP to the MEF usable EAP.

The usable EAP for the MRF-D EDL sourced by III MEF is still associated with III MEF supply accounts and thus incorporated in any III MEF S-rating calculation. Nevertheless, for illustrative purposes, for that EDL and for the USMC overall, we subtracted the EDL usable EAP. The implicit assumption here is that that EAP is not available for standard mission, but we do not mean to suggest that its use has been lost or it is without utility in its current role.

Table 2.5 shows the results. Eliminating the EDLs would raise the number of S-1 TAMCNs by 16, a change of 9 percent. The impact would be somewhat greater at I and II MEF. This impact, again, is illustrative. Provisional unit equipment, for instance, is in fact included in USMC overall usable EAP, so by the bounds of this analysis USMC overall S-ratings would not increase if the EDLs were eliminated. Nevertheless, it suggests the EDLs contain types and quantities of equipment that are significant relative to USMC supplies.

The qualitative analysis indicates that the EDL impact on Marine Corps readiness varies significantly by TAMCN and source. For some TAMCNs and sourcing pools, EDLs constitute over 25 percent of the AAO, including HD/LD items. But for most, their percentage is less. For almost half of reportable TAMCNs, the EDL requirement is smaller than the quantity by which equipment on hand across the Marine Corps exceeds the total AAO. The issue then may be more directly linked to management practices and information visibility issues.

Table 2.5
Illustrative Supply Impact of EDLs, October 2017

	S-1 TAMCNs	Notional Additional S-1 TAMCNs if EDLs Eliminated	Change to Total S-1 TAMCNs
Overall	169	16	9%
I MEF	150	22	15%
II MEF	135	20	15%
III MEF	152	10	7%

SOURCE: RAND analysis.

Gaps and Disconnects in Marine Corps Policies

The primary issue regarding Marine Corps policies and provisional units is that the way policies are currently written is not supportive of the needs of provisional units. These policies were created in an ad hoc manner and are not well understood by the force they intend to assist.

Marine Corps logistics (supply, maintenance, distribution, etc.) policies that account for provisional units for the most part focus on the establishment of the unit (EDL creation and validation) and readiness reporting of equipment. The primary gap in policy is what is required for the sustainment of these units as they become enduring requirements outside of the standard GFM and AAO processes. Another challenge is that many of the systems referenced in policy are not able to support the requirements of provisional units, which results in a disconnect between how accountability should work in theory and how it works in practice. Even with specified policy, some SMEs when interviewed were unaware that these policies existed, resulting in inconsistent application across the sourcing units and MARFORs. Finally, there is an underlying issue in that the way policy is currently written, provisional units are intended to be temporary in nature, but in practice they have become enduring rotational commitments. How to account for that disconnect in policy is a continued challenge.

A list of the policies reviewed for this project is presented in Table 2.6. Most of the references were provided by the sponsor, but additional

Table 2.6
Marine Corps Logistics Policies Reviewed

Marine Corps Policy	Date	Description	Reference to Provisional Units[a]
MCO 3000.11E Ground Equipment Condition and Supply Materiel Readiness Reporting (MRR) Policy	April 5, 2012	Outlines requirements for reporting readiness for ground equipment that pertain to its condition and inventory levels.	Yes. Provides clarifying guidance on how to report assets being used in support of task-organized units.
MCO 3000.13A Marine Corps Readiness Reporting	July 18, 2017	Provides policy and procedures to units, selected installations, and other organizations in the Marine Corps on reporting readiness to meet service and Department of Defense (DoD) reporting requirements.	Yes. Readiness reporting procedures for personnel (via manning document), equipment (via EDL), and training (via mission unit) is task organized to conduct.
MCO 3120.12 Marine Corps Global Force Management and Force Synchronization Manual	February 11, 2015	Outlines the process for assigning forces to the combatant commands in accordance with requirements for GFM and provides "policy, guidance and direction for the execution of Force Synchronization and Force Generation processes."	Limited
MCO 4400.150 E Consumer-Level Supply Policy Manual	January 29, 2014	Guidance for consumer-level supply operations.	Yes. Guidance on command adjustments and readiness reporting.

Table 2.6—Continued

Marine Corps Policy	Date	Description	Reference to Provisional Units[a]
MCO 4400.151B Intermediate-Level Supply Management Policy Manual	December 14, 2012	Establishes supply policies for intermediate-level supply accounts.	No
MCO 4400.201-V3 Management of Property in the Possession of the Marine Corps Volume 3, Chapter 2	June 13, 2016	Establishes supply policy for effective control of USMC resources.	Yes. Definitive policy document for equipping and sustaining task-organized provisional units. Provides policy, procedures, and guidance on how to form, equip, and sustain task-organized provisional units.
MCO 5311.1E Total Force Structure Process	November 18, 2015	Policy and procedural guidance on the Total Force Structure Process (TFSP) and Capabilities Based Assessment (CBA) Planning.	Yes. Specifically mentions task-organized equip policies and procedures that are in line with MCO 4400.201-V3 (e.g., EDL development, approval, and fulfillment processes and procedures).
UM 4000-125 Global Combat Support System-Marine Corps (GCSS-MC) User Manual	August 11, 2017	Provides user guidance on functional procedures for supply and maintenance functions in both garrison and deployed environments.	Yes

SOURCE: RAND analysis.

[a] Marine Corps policies refer to provisional units as "task-organized units."

materials were reviewed that were deemed relevant. The focus was on those specific policies that related to the equipping of provisional units.

Current Policies Do Not Address Sustainment of Provisional Units

Most of the policies reviewed focus on requirements for establishing a provisional unit and maintaining equipment accountability. Multiple gaps in policy were discovered during the course of analysis. The first gap was that sustainment issues were not addressed in policy. MCO 4400.201, which provides the most comprehensive guidance on equipping provisional units, provides ample guidance for how to establish a provisional unit's equipping requirements; however, when it comes to sustainment, the policy includes just one paragraph:

> *If a SPMAGTF or task organized unit uses a rotational force model, the supported Marine Force will hold an equipment review conference no later than 90 days following the relief-in-place/transfer of authority "RIP/TOA" to determine any necessary changes to the equipment needs list. If changes are required the review, validation, and approval process is the same as listed in the above steps, including the TFSMS [Total Force Structure Management System] actions.*[19]

While this guidance is useful in ensuring that requirements for follow-on rotations are validated and provided, it does not address other equipping challenges that units are facing as a result of the enduring rotational requirement. A primary example is there is no guidance on how provisional units will field updated or modernized equipment.[20] This is due in part because MCO 4400.201 severs the

---

[19] MCO 4400.201.

[20] An example is the recent fielding of new fixed headspace for .50 caliber machine guns. Fielding of the updated equipment was based on garrison infantry battalion authorizations—for example, forty total for an infantry regiment (ten per battalion). However, for the infantry battalion supporting the SPMAGTF, there is a requirement to support the provisional unit EDL, which was approved for thirty-five machine guns, but the infantry regiment is only provided ten upgrades for that battalion. The MARFOR therefore has to request the modernized equipment but relies on the sourcing regiment to advocate.

relationship between force provider and component once equipment is redistributed. This leads to life-cycle management issues where the force-providing MEFs do not accurately budget for sustainment requirements demanded by the provisional units.

Complementary to MCO 4400.201 is MCO 5311.1E, the TFSP. The document provides policy and procedural guidance on the TFSP and Capabilities Based Assessment (CBA) Planning. It specifically mentions task-organized equipment policies and procedures that are in line with MCO 4400.201 (e.g., EDL development, approval, and fulfillment processes and procedures). However, similar to MCO 4400.201, there are very few procedures that focus on the sustainment of these units as they became enduring rotational requirements. There is currently no mechanism to develop and track EDLs based on provisional unit concepts of employment. By design the Marine Corps has established that provisional unit EDLs will not generate a demand signal or register a requirement that increase equipment AAOs.[21] While it is understandable that the Marine Corps does not want AAOs to increase, by not capturing the equipment demand signals in a meaningful way, the Marine Corps loses visibility on how Marines are truly using equipment.

There are mechanisms outlined in the TFSP that try to establish visibility of equipment demand by provisional units. One of these is a portal within TFSMS for loading provisional unit EDLs. But, as will be discussed later, this portal is a dead end because it does not talk to other Marine Corps systems. Therefore, units interviewed were not consistent in loading their EDLs into TFSMS. Many units were providing EDL spreadsheets to I&L as their accountability process. Without systematic access to historical EDLs, they were available to be reviewed or to observe trends in equipment requested and utilized.

Another mechanism mentioned in MCO 5311.1E that would signal to HQMC an increase in demand for certain pieces of equipment is the Table of Organization and Equipment Change Requests (TOECRs). These requests are sent to the TFSD when a unit requires a

---

[21] MCO 5311.1E.

change to its authorization of personnel or equipment. However, when units were interviewed we found this process was rarely used because there is a perceived high likelihood of the request being rejected.

Information Systems Referenced in Policy Do Not Support Procedures as Written

The Ground Equipment Condition and Supply Materiel Readiness Reporting (MRR) Policy (MCO 3000.11E) outlines requirements for reporting readiness for ground equipment that pertain to its condition and inventory levels. Specific to provisional units, this order provides clarifying guidance on how to report assets being used in support of task-organized units. It provides clear guidance on how command adjustments are to be conducted (units that transfer equipment to provisional units still report that equipment against their TFSMS Table of Equipment [T/E] quantity). However, other challenges are present in the order.

The order discusses multiple information systems that assist in the accurate reporting of ground equipment readiness. It refers to Global Combat Support System-Marine Corps (GCSS-MC) as the program of record to report operational readiness particularly for deployed units. However, during interviews we found a disconnect between what GCSS-MC is supposed to provide information for and what it actually is able to accomplish. GCSS-MC is not accurately reflecting this operational readiness information for provisional units owing to the fact that the system only shows on-hand equipment levels, not EDL requirements. Therefore, all equipment looks like excess equipment on operational readiness reports. Because of these limitations with the system, units are not using the system as it is outlined.

Similarly, MCO 3000.11E establishes TFSMS as the authoritative data source that outlines a unit's T/E quantities. The order states that non-task-organized units will use T/E quantities from TFSMS to report operational readiness but task-organized units will report requirements based on what the unit is approved to use. Most provisional units, when interviewed, are not inputting their EDLs into TFSMS because the connection between TFSMS and GCSS-MC was severed for provisional units.[22] TFSMS is therefore a dead end for information reporting. Additionally, interviews with logistics officers at provisional units indicated that EDL approvals were not always timely. One unit was using an EDL that was already two years out of date, further deteriorating the accuracy of information being provided by Marine Corps programs of record.[23]

Are Provisional Units Temporary in Nature or Enduring Requirements?

There is a delicate balance between providing clear guidance but also allowing for flexibility in doctrinal policies. The tension is even more acute for units that are not intended to be enduring commitments. It also speaks to the broader disconnect between whether these units are temporary in nature as they were designed or, rather, enduring requirements. Most people interviewed agreed that the way that provisional units are currently employed is at odds with the way they are outlined in Marine Corps doctrine. As one interviewee articulated,

> We are supposed to be temporary. We are supposed to be short in duration—administratively not in compliance with what a SPMAGTF is. Until we decide if this is temporary or long term, that will feed TFSMS and UIC and MCC [monitored command code]. Those are tied into GCSS-MC. System doesn't see it as a real thing.

Many senior stakeholders echoed these sentiments, indicating that there may be a gap in current doctrinal definitions that fails to account for the fact that many provisional units have become enduring requirements. This contradicts MCO 4400.201 that specifies provisional units as "temporary" and "organized for a limited period of time."

---

[22] Interview data.

[23] Interview data.

IT Systems That Are Not Responsive to Provisional Unit and Enterprise Needs

We heard conflicting information from interviewees about how the readiness process and accountability processes worked. This indicates that current policies and procedures for accounting for provisional units using current IT systems are either difficult to follow, not well understood, or offer conflicting information.

Equipment Accountability

To maintain equipment accountability, the programs of record within the Marine Corps are TFSMS (to track authorized allowances) and GCSS-MC (to track on-hand quantities). While these systems and their associated processes for the most part work for standard Marine Corps units, the workarounds and adapted processes to use these systems for provisional units are cumbersome and fail to meet the unit's needs.[24] As a result, most units are defaulting to the use of spreadsheets to maintain visibility of on-hand equipment and requirements. One interviewee described the current system in the following way:

> We pull the MSR [material sustainment report] from GCSS and bounce off that; we can't enter anything into TFSMS, it's just an EDL spreadsheet. We don't have a TO&E. [We] select the consolidated memorandum receipt [CMR] from GCSS and validate the EDL, and those two combine to make up our MSR. Some of the reports are pulled from GCSS, [like] the CMR, but everything else is spreadsheets. Twice we have pulled the supply officers and gone line by line through the CMR and gone through the EDL spreadsheet.

Further complicating the process, there are difficulties with unit account managers retaining their permissions to IT systems while deployed in support of provisional units. Because these units are cobbled together from a variety of units, users have to be dropped from their parent command accounts and put on temporary assigned duty (TAD) so that the provisional unit user account manager can pick

---

[24] This information was provided during SME interviews.

them up. Oftentimes when the unit gets into theater these permissions are dropped with no notification because of inspections occurring back at home station.[25]

Readiness Reporting

Similarly, the Defense Readiness Reporting System (DRRS) is configured in a way that is supportive to the needs of standard Marine Corps units, but when it is used by provisional units it presents certain challenges. DRRS does not currently pull from TFSMS; therefore provisional units must manually enter their EDLs to report equipment readiness. Because units manually enter this information, an EDL will almost never have a supply score less than S-1 because EDLs are written based on what the unit possesses. For provisional units, the primary information provided by DRRS as it relates to equipment, then, is its condition, maintenance, and availability. EDLs are supposed to be an accurate picture of what each provisional unit possesses, but often they list what was supposed to be taken instead of what was actually deployed. This results in there sometimes being no record of equipment with a provisional unit until that equipment is broken.

Each SPMAGTF component—Command Element (CE), Ground Combat Element (GCE), Air (or Aviation) Combat Element (ACE),[26] Logistics Combat Element (LCE)—sources its gear independently and reports separate EDLs and manning documents in DRRS. For instance, the GCE and CE usually deploy around a standard unit formation such as an infantry battalion or infantry regiment headquarters, respectively. They then take additional augments, but over 50 percent of the unit belongs to the standard command. The LCE, on the other hand, is often comprised of detachments that are too small to require their own manning documents. This leads to different units reporting their readiness in inconsistent manners.[27]

---

[25] This information was provided during SME interviews.

[26] The ACE was excluded from the analysis at the sponsor's request.

[27] This information was provided during SME interviews.

Provisional unit components report readiness for their assigned missions in DRRS. The Command Element is the exception; it is assigned a core mission and not an additional assigned mission. Therefore, these components are tracked discretely using unique UICs, while other components are not.

Lack of Accurate Data on Utilization Rates

An additional challenge that was not identified at the beginning of the project but became evident as the project progressed was that none of the Marines interviewed were able to provide accurate assessments of utilization rates. Understanding utilization rates is critical for commanders to make informed decisions about what equipment is required or not required for the mission. This is closely tied with EDL validation and can lead to better distribution of resources across the Marine Corps.

Unit after-action reports indicated that some equipment was getting more utilization at certain locations as compared to others, but none of this information was captured in a systematic manner. For instance, we were informed that large pieces of equipment such as rolling stock owned by SPMAGTF-CR-CC was not being used for mission requirements. This was primarily because the equipment, such as Medium Tactical Vehicle Replacements (MTVRs), are not properly fitted with force protection requirements and so were restricted to use on base. Instead the unit was heavily utilizing Mine-Resistant Ambush Protected (MRAP) vehicles from the MAP-K pool. In the case of SPMAGTF-CR-AF, units are using equipment at varying rates. Equipment from Romania is frequently used, while the equipment at Naval Air Station Sigonella rarely accumulates mileage or hours.[28] Despite these differences in utilization rates it did not appear that units were incorporating these findings into their EDL validations to rotate gear back to home station.

[28] United States Marine Corps Centers for Lessons Learned, *Logistics Combat Element in Support of Special Purpose Marine Air Ground Task Force—Crisis Response—Africa 15.2 and 16.1 Rotation; Combat Logistics Battalion 6*, Washington, D.C., September 20, 2016.

Summary

While the challenges addressed in this chapter do not represent all of the issues faced by provisional units when it comes to equipping, they do represent the most pressing dilemmas facing the Marine Corps and provisional units. In summary, we analyzed the challenges presented and found that

- Tension exists between providing support to provisional units at the expense of GFM requirements.
- Provisional units are not factored into AAOs does stress the non-provisional force, but it should be noted that provisional unit demand does not exceed the full AAO for all TAMCNs. Only for some select TAMCNs is the demand relatively high relative to the sourcing MEF AAO.
- Policies adequately address how to establish provisional units but lack guidance on how to sustain units as they become enduring requirements. Issues with IT systems prevent policies from being implemented as they are written. Compliance and knowledge of policies within the operating force are uneven.
- IT systems that are stovepiped create significant hurdles, which require creative, ad hoc workarounds in order for units to maintain visibility of equipment and its readiness.
- Leaders are making EDL decisions without accurate assessments of utilization rates.

In Chapter Three, we discuss potential solution elements to address the issues raised in this chapter.

Potential Solutions for Equipping Challenges

In this chapter, we characterize potential solutions that might address the provisional equipping problems documented in Chapter Two. The conceptual approach that we took was to differentiate between *solution elements* that address discrete problems of provisional equipping and *courses of action* that comprehensively address multiple problems associated with provisional equipping. This distinction is a subtle but important one. Solution elements are single solutions for single problems. Courses of action are several solution elements that follow a unified strategic approach to addressing equipping problems. This enables Marines and civilians to have a single, clear understanding of the equipping process and the goals that it is trying to fulfill.

We first discuss the sources of potential solution elements and our process for identifying them. We then describe the solution elements themselves. Last, we discuss how groups of solution elements form potential courses of action. Those courses of action will then be evaluated in Chapter Four.

Sourcing Potential Solution Elements

Potential solutions to current equipping challenges described in Chapter Two arose from a variety of sources. Previous RAND research on Army and Marine Corps equipping policies and practices yielded several potential solutions, particularly those focused on economizing the use and maintenance of equipment that the enterprise already possesses.[1] While some of these solutions were not developed with the current context (i.e., provisional unit equipping) in mind, they still offer a variety of useful insights that have been analytically tested and, in some cases, employed.

In other cases, we inferred solutions from the problems that were described by interviewees. For example, if an interviewee identified the inability to shift funds between different elements of the AAO, then we would infer that a flexible AAO policy could be a solution element to be considered. A key advantage of inferred solutions is that they allowed us to reconsider potential solutions that interviewees either did not consider or decided were too infeasible to be suggested outright.

However, the greatest number of solution elements were directly offered by interviewees. In describing an element of the provisional unit equipping problem, interviewees frequently recommended unsolicited solutions. In other cases, interviewees were asked to assess a proposed set of solution elements. In both cases, solution elements sourced from those Marines who are most intimately familiar with the provisional unit equipping problem offered the most incisive ideas, with the added benefit of being specifically fitted to address the current provisional unit equipping problem.

A key challenge, however, was ensuring that we had a complete set of solutions for evaluation. Ideally, solution sets would be developed from analytical frameworks that could guarantee that the process delivered a mutually exclusive, collectively exhaustive set of solution elements. Because of the considerable variation in types and scale of provisional equipping challenges, we were not able to meet such a standard.

However, we used two techniques to provide confidence that the solution elements identified represented many, if not most, feasible solutions. The first technique was confining the problems to be validated and addressed to those initially identified by the sponsor.

---

[1] See Christopher G. Pernin et al., *Efficiencies from Applying a Rotational Equipping Strategy,* Santa Monica, Calif.: RAND Corporation, MG-1092-A, 2011; and Matthew W. Lewis et al., *New Equipping Strategies for Combat Support Hospitals,* Santa Monica, Calif.: RAND Corporation, MG-887-A, 2010.

This analytical framework ensured that the solutions defined were focused around the problem at hand. Second, we structured the interview process such that solution elements could be presented to a set of increasingly knowledgeable and experienced experts on provisional unit equipping. For example, the team conducted interviews with provisional unit logistics representatives before interviewing (and presenting solutions to) enterprise-level experts. This allowed us to refine the solution set iteratively. The result is a large and robust set of solution elements for evaluation.

Solution Elements Identified

Solution elements proposed by interviewees, culled from previous research, or inferred by us are listed in Table 3.1.

Greater Enterprise Visibility

A common challenge cited by interviewees was the difficulty of maintaining equipment visibility between provisional units and the parent units that their equipment is sourced from. According to MCO 4400.201,

> *MARFORS and the MEFs will identify equipment sourced to task-organized forces during MARCORLOGCOM's quarterly push equipment sourcing report/process. This serves an internal control to ensure that MARCORLOGCOM and the MARFORs have visibility of equipment sourced to EDLs.*

This lack of visibility stems from uneven adherence to policies on how provisional unit equipping should be accounted for and, consequently, the inability of IT systems to support that accountability. Upgrading, refreshing, or investing in new IT solutions would be a positive step toward maintaining accountability of equipment.

Table 3.1
Solution Element Table

Solution Element	Advantages	Disadvantages
Greater enterprise visibility	Provides truer insight into nature of equipping problem.	Does not alleviate equipping shortages.
AAO flexibility	More responsive, economical alleviation of equipment burden.	Unclear maintenance responsibility. Regular usage increases operational risk.
Forward-deployed equipment sets	Allows USMC to take on missions without increasing operational risk. Relieves MEF/MARFOR equipping burden.	Costly. Changing missions may not be suitable for static equipment set.
HD/LD considerations	Buying to specific HD/LD requirements is more economical than buying full provisional unit equipping sets.	Not all HD/LD equipment can be purchased owing to closed production lines or other factors. For electronics (a large percentage of HD/LD equipment), technology becomes rapidly obsolete.
Provisional unit-specific policies	Efficiency gains through consistent policies followed by all. Some policies already exist as practices and only need codification.	Does not address HD/LD issues. New policies may have second or third-order effects.
Right-size maintainer force	Ensures sustainable and responsible use of any equipment sourced to provisional units.	Not likely to be prioritized. Does not address equipment shortage problem.
Reserve maintenance augmentation	Economical alternative to right-sizing maintainer force.	Increased operational risk if large-scale contingency develops.
Joint/partner maintenance support	Economizes on available, sometimes underused assets.	Technical challenges to maintaining USMC equipment. Parts ordering accountability (especially for USMC parts) may be difficult.

Table 3.1—Continued

Solution Element	Advantages	Disadvantages
Stricter validation of provisional unit missions	Makes equipment considerations explicit to commander. Reduces mission creep.	May increase operational risk. Lower utilization rate for units.

SOURCE: RAND analysis.

Flexibility in AAO Pillars

Some provisional units are stationed or deployed near existing equipment sets that are allocated to the war reserve program (colloquially known as one of several "pillars") of the AAO. These programs include MPF sets, MCPP-N, the MAP-K, and the War Reserve Materiel Requirements (WRMR). While some provisional units have established agreements to draw equipment from MAP-K, other units indicated that drawing equipment from war reserve pillars for contingency operations is either not allowed or difficult to accomplish.[2] Updating the war reserve policy to make accessing these pillars of the AAO possible would help alleviate the sourcing burden for MEFs and MARFORs and possibly reduce transportation costs. An updated policy would specify what kinds of equipment could be drawn, duration, return policies, and maintenance burden-sharing between the provisional unit, MARCORLOGCOM, and the war reserve caretakers.

Forward-Deployed Equipment Sets

We inferred from interviews and previous research on tailored equipping options that equipment pools that are forward deployed at concentration sites could alleviate the burden of equipment sourcing on the MEFs and MARFORs. These equipment sets would be added to the AAO and used specifically to support provisional units.

Additional Procurement for Select Items

We reviewed provisional unit EDLs and cost data and found that while provisional unit EDLs did not seem to place undue stress on most TAMCNs, they accounted for a large share of demand for some.[3] Some of those were on the USMC's official HD/LD list. In lieu of increasing the AAO for all provisional equipment, the Marine Corps could selectively procure additional quantities of a more limited set of items—those on the HD/LD list and perhaps others as deemed warranted—to alleviate the demand. Delays in accomplishing additional equipment

---

[2] Summarized from SME interviews.

[3] See Chapter Two for a more detailed discussion of HD/LD TAMCNs.

procurement (e.g., closed production lines) and the projected longevity of the specific requirement should be considered in such cases.

Inclusion of Provisional Units in Existing Policies

Generally, interviewees found little consideration for provisional units in existing force development, equipping, and reporting policies. While the concept of SPMAGTFs and provisional units has long been established, the Marine Corps did not foresee the multiyear duration of the current group of provisional units. The Marine Corps should reconsider existing policies to account for the possibility of long-term employment of such units. In some cases, new policies that result in second- and third-order effects need not be written. The Marine Corps will merely need to choose between one of several existing practices (e.g., equipment sourcing, excess equipment management) to ensure consistency. In other cases, new policies (e.g., equipment visibility) will need to be developed.

Rebalance the Maintenance Force

Several interviewees identified the growing gap between the size of the Marine Corps' growing maintenance burden (in terms of number, variety, and complexity of ground equipment) and the relative stagnation of the size and experience level of the maintenance force. While interviewees noted that this is a Marine Corps–wide problem, provisional units are particularly vulnerable to the gap's effects because of commanders' ability and tendency to staff provisional units with more operators than maintainers.

Interviewees recommended an enterprise-wide rebalance between the size of the maintenance force and the ground equipment inventory. Interviewees believed that part of this rebalance would move some home station maintenance functions to more cost- and performance-effective government civilian workers, allowing the military maintainers to focus on maintenance in more operational or expeditionary environments. Furthermore, we identified an experience gap that may require a larger maintenance force overall. It will also require a more non-commissioned officer (NCO) and staff non-commissioned officer (SNCO)-heavy maintenance force structure to ensure provisional units have maintenance personnel with enough time- and experience-based knowledge to address provisional units' maintenance needs without the benefit of a parent unit to provide support.

Reserve Maintenance Augmentation

We identified the reserve component maintenance force as an economical alternative to increasing the civilian maintenance workforce or the active component maintenance force. Reserve component maintenance personnel can augment provisional units as part of active duty special work (ADSW) or other mobilization programs. A key disadvantage would be increased operational risk if a new contingency (that requires reserve mobilizations) strained the reserve component's maintenance capacity.

Joint and Partner Maintenance Support

Interviewees suggested using joint and partner maintenance capabilities during provisional unit deployments as a supplement to organic capabilities. A key challenge to this would be that the Marine Corps and its partners would need to develop implementation guidelines, such as parts-ordering mechanisms and responsibilities, repair knowledge sharing, and others. Another challenge is that some ground equipment and its components might be so unique to the Marine Corps that joint and partner maintenance capabilities may not be sufficient to meet the provisional unit's needs. However, interviewees cited several existing and successful practices of relying on Army maintenance capabilities (for armor) and partner maintenance capabilities for tire disposal and corrosion control for ground vehicles generally.

Stricter Validation of Provisional Unit Missions

Several senior interviewees identified mission creep as a key driver of the provisional unit equipping problem. Although some provisional units were formed for specific purposes, some units have exceeded their original missions or have no obvious end point. Interviewees stated that when missions are defined without discrete limits, provisional unit commanders are more apt to take on additional missions in order to increase the perceived purpose for their units. This in turn drives up equipment requirements to meet new mission needs.

Interviewees suggested that commanders take greater care to understand the maintenance burden and implications of taking on new missions that require additional equipment. Higher-level head-quarters regimental, division, and MEF can scrutinize requests more. Last, HQMC, PP&O could implement stricter equipment requirement validation processes to further discipline the volume of provisional unit equipment requests.

Developing Courses of Action from Solution Elements

When considering solution elements, some similarities were clearly apparent between clusters of elements. By identifying and expanding on common threads within solution elements, we developed courses of action with coherent strategic focuses. This approach allows policymakers to explicitly consider and better prioritize their choices. The courses of action created from these solution elements include: policy implementation, centralized management, permanent structure, tailored equipping, and leasing or renting.

Policy Implementation
The policy implementation course of action combines these solution elements:

- enterprise visibility
- provisional unit–specific policies
- stricter validation of provisional unit missions.

These solution elements would provide consistent execution standards for all provisional units. The first step in this course of action is to ensure that regulations, from Marine Corps orders to individual unit policies and standard operating procedures, are aligned. These regulations should include those governing EDL validation, EDL visibility in TFSMS, accurate DRRS reporting, and consistent GCSS-MC accounting. One further step would be an updating of these resource management platforms with provisional unit–specific functionality.

Centralized Management of Excess

The centralized management course of action combines these solution elements:

- enterprise visibility
- HD/LD considerations.

In this course of action, provisional equipment is managed centrally rather than at the MEFs. In our example, MARCORLOG-COM would be the central manager, managing equipment in excess of the AAO across the Marine Corps and converting it for provisional unit use. MARCORLOGCOM would create an operational reserve float (ORF) with unique supply codes separate from other managed inventory.

MARCORLOGCOM would be responsible for readiness and sustainability of the equipment, relieving the MARFORs that are currently responsible. The scope of this ORF would be sizable equipment that is larger than a High Mobility Multipurpose Wheeled Vehicle (HMMWV) and HD/LD items such as very small aperture terminals (VSATs). All other equipment would remain with its present owner. The aim of this centralization process is to acknowledge that excess equipment is the main source of provisional unit equipping and that a consistent approach to accountability and maintenance will alleviate issues currently associated with provisional unit equipping.

Permanent Structure

The permanent structure course of action combines these solution elements:

- HD/LD considerations
- provisional unit–specific policies
- rightsizing maintainer force.

This course of action creates standing Tables of Equipment (T/Es) for provisional units. T/Es would be filled with excess equipment currently used by provisional units. More equipment would be procured for TAMCNs not filled by excess, which would primarily be for HD/LD equipment. An associated, nonequipping action would be to create standing Tables of Organization to man (and maintain) the equipment.

Tailored Equipping (Light, Medium, Heavy)

The tailored equipping course of action combines these solution elements:

- AAO flexibility
- forward-deployed equipment sets.

This course of action would concentrate (to varying degrees) all Marine Corps ground equipment around shared home station training sets, exercise sets in CONUS and around the world, and operational sets aligned to MARFORs. This would decrease home station equipment sets to fit unit training METs, provide baseline EDL or baseline-plus sets at training locations—for example, Marine Corps Air Ground Combat Center (MCAGCC) 29 Palms, California, and Camp Fuji—and operational sets tailored for MARFOR needs. Units would continue to maintain personal weapons, optics, communications equipment, and other low weight/cube equipment. A variant of this course of action would include converting war reserve elements of the AAO into parts of the exercise or operational equipment sets.

Leasing/Renting

The leasing/renting course of action combines these solution elements:

- joint/partner maintenance support
- reserve maintenance augmentation
- HD/LD considerations.

The leasing/renting option seeks to economize on provisional unit equipping by using non–Marine Corps resources to conduct mainte-

Table 3.2
Relationship Between Solution Elements and Courses of Action

	Policy Implementation	Centralized Management	Permanent Structure	Tailored Equipping	Leasing/Renting
Greater enterprise visibility	x	x			
AAO flexibility				x	
Forward-deployed equipment sets				x	
HD/LD considerations		x	x		x
Provisional unit–specific policies	x		x		
Rightsize maintainer force			x		
Reserve maintenance augmentation					x
Joint/partner maintenance support					x
Stricter mission analysis	x				

SOURCE: RAND analysis.

nance for provisional units as well as leasing or renting some equipment, such as generators, with commercially available equivalents. A combination of such initiatives can help alleviate some of the supply and sustainment issues associated with provisional unit equipping.

Each of the COAs outlined above capitalize on certain solution elements. These elements by COA are summarized in Table 3.2.

Summary

The solution space for provisional unit equipping issues must balance the need to be effective in very specific instances with the demand for generalizability across organizations. This is for two reasons. First, provisional unit equipping is a complex process with a variety of stakeholders across non-alike organizations and echelons. Second, bespoke solutions for each individual variant of the overall problem would however prevent the Marine Corps from achieving the flexibility and efficiency required to address new forms of provisional units and their equipping needs in the future.

Because of this, we have taken care to take both bottom-up and top-down approaches to developing policy recommendations. In this chapter, we identified the discrete solution elements that we have found through previous research or interviewing provisional unit equipping stakeholders. To balance the precise solution elements with the need for generalizable solutions, we consolidated similar solution elements into more comprehensive courses of action. In Chapter Four, we take the next step toward generalizing our recommendations by assessing the costs, benefits, and risks of each course of action.

Assessing Alternate Equipping Strategies

Building on the solution elements outlined in Chapter Three, we now will discuss which courses of action were chosen for analysis and why. These in-depth descriptions will include strengths, weaknesses, and cost discussions for each of the courses of actions (COAs). The four COAs evaluated are

- policy implementation
- centralized management
- permanent structure
- tailored equipping.

Choosing Courses of Action for Analysis

As outlined in Chapter Three, courses of action were developed by clustering similar solution elements into meaningful strategies for addressing equipping challenges. These ranged from little change to the status quo to significant changes to current Marine Corps equipping strategies.

To determine which solution strategies to analyze, the team took guidance from the Marine Corps planning process that requires COAs to be feasible, acceptable, suitable, distinguishable, and complete.[1] Using this guidance and input from key stakeholders, we determined

---

[1] Headquarters Marine Corps, *Marine Corps Planning Process*, Marine Corps Warfighting Publication 5-1, Washington, D.C., August 24, 2010.

that the most distinguishable COAs were to implement the status quo, with selected changes and required adherence to existing policies; make provisional units part of the permanent Marine Corps structure; centralize management at MARCORLOGCOM; and pursue some form of tailored equipping.

Another course of action that was considered but excluded was to lease or contract high-demand items. This strategy did not constitute a complete and fully formed COA as it was limited to addressing a subset of challenges that were of lower priority to key decisionmakers. Discussion with key stakeholders resulted in the elimination of this strategy. Focus remained on those strategies that combined the most solution elements in a complete and distinguishable manner. Table 4.1 summarizes the four COAs analyzed.

Each COA was evaluated for strengths and weaknesses based on subject matter input and team analysis. Additionally, we worked with key stakeholders to develop cost estimates with enough fidelity to adequately ensure discrimination between course of action. The deliverables as a result of the cost analysis will include a rough order of magnitude for annual costs to the Marine Corps. Specific costs, including any personnel, facilities, and maintenance costs, for each of the alternatives are discussed further in Appendix C.

Policy Implementation

This strategy was designed to specifically address accountability and visibility challenges identified. The key policy requiring implementation is the MCO 4400.201. This order outlines processes for validating EDLs, loading EDLs in TFSMS, and accurately reporting on-hand equipment in DRRS and GCSS-MC. While there are portions of the order that would benefit from being rewritten or clarified, for the most part, better implementation of policies outlined in this order would be advantageous to the Marine Corps.

Table 4.1
Alternate Equipping Strategies

Equipping Strategy	Summary
Policy Implementation	• Ensure validation of EDLs for each rotation, load EDLs in TFSMS, do accurate DRRS-MC reporting and appropriate GCSS-MC accounting.

Centralized Management	• MARCORLOGCOM manages excess equipment allowance pool:
	  • Create an ORF with unique supply code separate from MARCORLOGCOM inventory.
	  • Gear is left at MARCORLOGCOM; does not affect MEF TEs.
	  • MARCORLOGCOM is responsible for readiness and sustainability, not MARFOR.
	• Scope: large equipment (>HMMWV) and HD/LD items (e.g., SATCOM); all other equipment remains with unit.

Permanent Structure	• Create standing Tables of Organization and Equipment:
	  • Assign excess equipment a new TO&E (unfunded).
	  • Procure additional equipment not filled by excess—primarily for HD/LD (funded).

Tailored Equipping	• Decrease home station equipment sets to fit unit training METs. • Continue to maintain all personal weapons, optics, comms, low weight/cube equipment. • Training sets: baseline EDL or baseline-plus at training locations (29 Palms, CA, III MEF). • Operating sets: stage tailored at MARFORs; in line with MARCORLOGCOM strategic plan to create forward-positioned global storage facilities; maintained by contractors/partners. • Proposed idea: convert war reserves to EAP like entities.

SOURCE: RAND analysis.

Advantages/Disadvantages

There are several key advantages to this course of action. Visibility would be increased, possibly alleviating a significant amount of the provisional unit equipping problem outright. It also provides a way for the Marine Corps to reset and reassess its provisional unit equipping practices after several years of constantly changing provisional unit operations. Last, it costs nearly nothing to implement.

However, there are disadvantages to this course of action. Current policies may not provide correct solutions to the provisional equipping problem. Thus, it might only support better management of existing equipment and does not change either supply or demand.

Centralized Management

The centralized management course of action was developed to address challenges expressed by units tasked with supporting provisional units. Current provisional unit equipping policies put a large burden on the MARFORs and supporting MEFs when it comes to sourcing and maintaining equipment. To alleviate that burden, this strategy would shift the sourcing and maintenance burden to a centralized manager such as MARCORLOGCOM. Additionally, this course of action creates a centralized manager for excess equipment, ensuring better distribution of equipment across the enterprise. This would be accomplished by creating an ORF with a unique supply code separate from MARCORLOGCOM inventory.

Advantages and Disadvantages

There are several advantages to centralized management. Higher levels of continuity and consistency for provisional unit equipment would be attained. It would satisfy the need for an excess equipment manager at the enterprise level. As previously mentioned, it would also reduce the maintenance burden on the MEFs and MARFORs.

However, this course of action has several weaknesses. To begin with, it assumes that there is enough excess inventory to sustain an ORF and that the type and mix of this excess would meet the demands of provisional units. Discussions with SMEs indicated this most likely would not be the case for some equipment, particularly the HD/LD equipment already identified as well as the most up-to-date variations of certain equipment such as the modified MTVRs. For example, when III MEF attempted to source an artillery battery for MRF-D from excess Marine Corps equipment, it was only able to source 50 percent of it.[2] Furthermore, this course of action alters the supporting and sup-

---

2   The example of sourcing an artillery battery for MRF-D is a unique case but still illustrates a key challenge with sourcing provisional unit equipment requirements. The case is unique in that MRF-D is a provisional requirement that did not come from a geographic combatant command request for forces; instead MRF-D is the result of a presidential directive. As such, growth in requirements for MRF-D has had little HQMC guidance and has been at the discretion of the MARFORPAC commander. III MEF directly sources MRF-D

ported unit relationship between the provisional unit and its higher headquarters. As a corollary, lower-priority provisional units would lose their dedicated MARFOR advocate. Finally, there would need to be an increase in personnel at MARCORLOGCOM to manage the requirements. MARCORLOGCOM also expressed concern that it is not in a position, nor does it have the expertise, to tell Marine Corps headquarters where it would need to accept risk regarding units.

Costs

This was assessed to be a relatively low-cost COA as compared to the others. There would be costs associated with a small increase in personnel at LOGCOM and funding to ensure LOGCOM is adequately resourced to implement this COA. There is potential for an uptick in time-based maintenance requirements, but there are no other anticipated large maintenance costs associated with this COA. Instead, the responsibility for those costs would shift. Finally, this COA requires no additional acquisition or facilities costs.

Permanent Structure

This strategy was one that was often raised and discarded in discussions with SMEs. Similar to the recent permanence established for MEU Combat Logistics Battalions (CLBs), this COA would take the EDLs for provisional units and make them permanent T/Es. Sourcing of equipment would be a combination of unfunded assignment using excess equipment and all other equipment being funded.

---

equipment requirements and as the force grows, its ability to source full requirements has been strained. For the artillery battery, a list of equipment was created based on the T/E of an artillery battery, reinforced. This list included 79 unique TAMCNs. MARFORPAC originally sourced what it could of these requirements from within the Division and III MEF, and then required global sourcing and was only able to fulfill 50 percent of the requirements. This illustrates how centralized management cannot fix the issue if the equipment is not there to be sourced to begin with. If the provisional unit requirement is outside of the AAO, the requirement for the additional artillery battery has not generated a demand signal.

Advantages and Disadvantages

The clear advantage of this course of action is that by establishing a permanent structure, units would no longer be forced to source equipment for rotational requirements. This permanency would ensure that sourcing of equipment would be done in a timely manner, which would eliminate other issues related to modernization and accountability. It would adhere to the standard equipping framework referenced in Marine Corps policy used by the Marine Corps and understood by Marines. Furthermore, it would provide a buffer for long-duration or high-intensity contingencies.

This course of action is not without drawbacks, however. Cost would be a major disadvantage; the procurement, operations, and maintenance (especially at the depot level) costs of the new equipment would be significantly greater than other options. It would also have secondary effects, such as either increasing force structure overall or the reassignment of personnel. Perhaps the biggest drawback, though, is that it commits the Marine Corps to a permanent structure, which limits flexibility to respond to other future crises.

Costs

This COA was assessed to be costlier than the centralized management strategy. It would require an increase in funding to transport all acquired equipment to the appropriate locations after initial procurement. The increase in size of T/Es and EDLs would also result in an increase in unit transportation costs. Similarly, more equipment would mean increased usage, which would result in increased maintenance costs. We assessed the cost of acquisition to a full complement, even with the redistribution of equipment, to be approximately $200 million.

Tailored Equipping

Based on other work done by RAND,[3] a tailored equipping strategy was proposed as a potential course of action for the Marine Corps. This strategy would change the way the Marine Corps equips units across the enterprise and have impacts on units other than just the provisional units. This COA would decrease home unit sets, establish additional training sets at training locations other than 29 Palms, CA and create forward-staged equipping sets within the areas of responsibility of the MARFORs. These sets could be similar to MAP-K or could capitalize on already forward-positioned sets of equipment such as MCPP-N, MAP-K, and Maritime Prepositioning Ships (MPS).

Advantages and Disadvantages

This course of action has several distinct advantages. Primarily, it is operationally responsive. Current experiences with MAP-K indicate that equipment would be easy to draw and maintain. It would also reduce transportation time for deployed provisional units, with units either drawing from MARFOR-owned equipment sets, war reserve materiel, or both. Even with difficulties moving equipment in theater because of country-specific requirements, it is estimated that the average transportation time would be reduced, resulting in better responsiveness. Across the enterprise, the Marine Corps could divest itself of excess equipment (and its attendant costs of ownership) and remaining equipment sets would be used at a higher rate.

However, this course of action is clearly the most difficult to implement and comes with a set of unique challenges. Operational risk is heightened if equipment from war reserve stocks is needed for additional contingencies. The skills of uniformed maintainers would atrophy as they focus less and less on maintenance during predeployment training at home station. Finally, this course of action would entail a significant cultural change. Commanders would need to be comfortable not "owning" the equipment. Troops would need to con-

---

[3]   Lewis et al., *New Equipping Strategies for Combat Support Hospitals*, 2010.

tinue to display a sense of ownership in the equipment that may not come naturally.

Costs

While this COA would be costly up front to efficiently rightsize the force, cost savings would be achieved in the long term. To establish training pools, an initial increase in personnel and facilities will need to be funded. However, transportation costs would decrease in the long run as equipment will be where it is needed when it is needed. Long-term acquisition savings can also be achieved as the equipment is right-sized to the force.

Table 4.2 summarizes the strengths and weaknesses of each of the courses of action analyzed.

Hybrid Approach

We asked interviewees for their perspective on each of the courses of action. The transcripts were analyzed to glean strengths and weaknesses as well as to capture whether stakeholders were for, against, or neutral on each of the strategies. The strengths and weaknesses articulated during those interviews are included in Table 4.2. As for interviewees' preference for one course of action over another, different stakeholders had different preferences. Operators within the provisional units themselves showed a slight preference for the permanent structure course of action. They felt that this COA would provide them with the gear they needed, when they needed it. Other stakeholders, particularly those at the HQMC level, showed high levels of preference for the tailored equipping and centralized management COAs. While they articulated weaknesses of each, stakeholders saw significant merit in both strategies. Therefore, we recommend a hybrid approach that capitalizes off the strengths of each COA.

A hybrid approach taking the most popular aspects of each of the COAs would be to have a centralized management function at MAR-CORLOGCOM to manage excess equipment inventory and remove the burden of redistribution from the MEFs. To address the HD/LD

Table 4.2
Strengths and Weaknesses of Alternate Equipping Strategies

Equipping Strategy	Strengths	Weaknesses
Policy Implementation	• Increased visibility. • Not resource intensive.	• Current policies may not be correct solution to provisional unit issues. • Does not address modernization issues or HD/LD deficiencies. • Historical lack of implementation and enforcement.
Centralized Management	• Reduces burden on MEFs/MARFORs.	• Assumes units want what is in MARCORLOGCOM inventory. • MARCORLOGCOM is not focused on operational risk.
Permanent Structure	• Stability. • Less risk to provisional unit.	• High cost (depot, O&M). • Lack of flexibility.
Tailored Equipping	• Operationally focused. • Responsive.	• Operational risk. • Requires paradigm shift. • Rightsizing units is costly and difficult.

SOURCE: RAND analysis.
SOURCE: Dedoose analysis of SME interviews.

shortfalls, something similar to the permanent structure COA would be recommended, but in lieu of increasing the AAO for all provisional equipment, the Marine Corps could selectively procure additional HD/LD items to alleviate the demand. Finally, the Marine Corps already conducts aspects of the tailored equipping COA by placing equipment sets in forward-deployed locations (MAP-K, MCPP-N, MPS). Making it easier to pull from these pillars of the AAO could assist in alleviating some problems faced by provisional units.

Summary

After evaluation of the four equipping strategies, no single course of action emerged as solving the majority of the equipping challenges identified. Therefore, a hybrid approach that capitalizes on the strengths of each COA is recommended. Costing for the COAs was difficult to do given the availability of data. It is recommended that the Marine Corps collect the following data to effectively cost-analyze a hybrid COA.

- Develop accurate maintenance cost records of dollars spent in relation to unit operational tempo:
  - Gather actual expenditures from all provisional units for the past three years to create an average maintenance cost per unit
  - Gather operational tempo information to include utilization rates
  - Determine whether equipment is in better or worse condition than similar equipment residing in CONUS and adjust estimates accordingly
- Gather cube and weight to get improved United States Transportation Command (USTRANSCOM) transportation estimates
- Find an analogous organization on which to base staffing assumptions.

CHAPTER FIVE

Recommendations

RAND was tasked to identify provisional unit equipping challenges, develop and analyze costs and benefits of alternative strategies, and provide recommendations. In this chapter, we articulate our final recommendation that the Marine Corps move forward with a multi-pronged strategy to fully address the provisional unit equipping issue. We believe this strategy should include

- risk matrices to prioritize unit equipping in the GFM process
- more empirical assessments of unit mission and requirements and improved collection of utilization rates
- updated Marine Corps policies that account for provisional and rotational unit equipping
- forward-positioned, flexible equipping strategies
- improved communication and visibility across IT systems
- elimination of the PEI rotation policy.

The specific recommendations are informed by three principles: an imperative to balance provisional unit equipping with overall readiness, a desire to minimize disruption to current Marine Corps practices, and an instinct to accommodate provisional unit equipping needs while keeping policies and practices flexible enough to accommodate future needs.

Use of Risk Matrices to Prioritize Unit Equipping

When discussing provisional unit equipping, many senior leaders asked us to consider the effect of prioritizing provisional unit equipping on the Marine Corps' overall readiness to react to major contingencies. The underlying tension is that if the Marine Corps sources units, or unit elements, for provisional or rotational missions, then these units (and their equipment) will not be available for major contingencies.

Addressing that tension is outside the scope of this project, but a more explicit articulation of the tension between current provisional missions and possible future contingencies can be useful in the Global Force Management process. One suggestion is to develop risk matrices that explicitly characterize the trade-off between fully supporting provisional missions and retaining forces in readiness for potential contingencies. As articulated by an interviewee:

> The Marine Corps should work on some kind of graduated risk matrix for equipment, and bin equipment from high, mid-, to lowend, to make sure high-end items aren't being sent to low-threat environments, like crisis response and humanitarian [assistance operations].

The primary characteristics that should be made explicit are the amount of warning time expected, the estimated scale of operations, threat level, potential duration of engagement, and mission complexity. Table 5.1 illustrates this approach. Those characteristics on the top row would result in higher priority for higher-end equipment and redistribution of assets, while those units demonstrating the characteristics on the bottom row would use other low-end equipping solutions.

Therefore, a short-term contingency in a low-threat environment could rely more heavily on low-end or commercial equipment solutions. An example would be SPMAGTF-CR-AF, which already relies heavily on commercial sourcing of equipment such as generators and transportation equipment. By comparison, units operating in high-threat environments and conducting complex mission profiles would be a priority for high-end equipment. An example of this type of unit would be SPMAGTF-CR-CC, which operates in a high-threat environment, with complex mission sets lasting of a longer duration, and

Table 5.1
Contingency Characteristics

Warning	Scale	Threat	Duration	Mission
None	Large	High	Long	Complex
Ample	Small	Low	Short	Simple

SOURCE: RAND analysis.

would therefore be the first priority for HD/LD assets that may be in limited supply and other force protection gear. Again, this is something the Marine Corps already does to a certain extent, but the decision-making is not clearly conveyed to the force.

By using a risk assessment approach to equipment planning, the Marine Corps can be more explicit about some organizations fulfilling economy-of-force missions. If this is the case, the starting point for equipping would be to start with excess Marine Corps equipment and then use alternative sourcing, such as buying or leasing commercial off-the-shelf (COTS) items. The lowest priority would be to source requirements from other low-priority units.

Additional Empirical Assessment of Unit Mission and Requirements

In addition to a more explicit understanding of risk in prioritizing provisional missions, the Marine Corps should also scrutinize provisional mission equipment requirements in greater detail. As noted in Chapter Two, the current assessment process (i.e., EDL validation) is less effective because of the subjective nature of these validations. More empirical assessments using data such as odometer readings, maintenance status, and other usage-rate metrics can, over time, provide the Marine Corps with an increasingly accurate picture of what equipment is actually needed by provisional units. Analyses of equipment usage data could also inform a commander's understanding of the value that EDL equipment, by TAMCN, provided to past commanders. For example, even moderately accurate odometer readings for vehicles, recorded regularly, over time, could provide insights on what vehicles were used, how often, and when. An example of such research is a RAND study that focused on providing unit commanders detailed data on how their equipment sets are used across the Army force generation training cycle: during reset, predeployment training, during deployments, and the after redeployment. In 2011 Arroyo Center research on support implications for new materiel management concepts, a case study of odometer reading data from a Stryker Brigade Combat Team, revealed large variations in the miles driven by individual vehicles at different stages of the deployment cycle. Commanders also learned that up to 50 percent of their Strykers had little or no use for many months during predeployment, as well as during the deployment. And unsurprisingly, there were surges in usage during predeployment training exercises.

Variations to an EDL baseline, established via analyses, must be justified in detail. This does not detract from the priority that commanders on the ground rightly expect; rather, it balances their needs with the needs of other units and missions. Once data sources and collection mechanisms are in place, big data tools from industry could also help the enterprise better understand equipment utilization. These tools could help decisionmakers visualize frequency of use, dependencies, rhythms, hot spots, and overload points. However, these tools are only a secondary step as it is essential that good data collection practices be implemented first. Once those collection methods are in place the proper visualization and analysis tools can be selected.

The Marine Corps should also carefully monitor the changing scope of missions that provisional units undertake. As noted in Chapter Two, most provisional units have significantly expanded the scope of their missions since the first rotation of their units. Since the Marine Corps is rightly attentive to commanders' needs, this exacerbates equipment demand. Stricter mission validation will alleviate this situation, but at the expense of lower utility to combatant commanders. Nevertheless, a more rigorous and continuous mission evaluation process that includes all stakeholders (e.g., combatant commanders, provisional units, sourcing units, and Headquarters Marine Corps) is warranted.[1]

---

[1]   Two recommended opportunities for this evaluation process would be at the Marine Corps operations summit or during the force synchronization process.

Updated Marine Corps Policies to Address Provisional Unit Sustainment and Reflect Current Capabilities of IT Systems

Another challenge to a more efficient and effective provisional unit equipping process is that existing Marine Corps policy does not provide adequate guidance to meet provisional units' unique equipping needs, nor does if reflect current practices. Although broad policies support flexibility, the seemingly enduring nature of provisional unit equipping warrants a more detailed articulation of processes, considerations, and end states to alleviate current challenges. As a starting point, the Marine Corps should address sustainment issues within MCO 4400.201 to include guidance on how modernization and updated equipment will be sourced to provisional units. A more detailed description of the supported and supporting relationship between provisional units and the MARFORs and MEFs should be better clarified with clear outlines of roles and responsibilities. Additionally, orders and users manuals should be updated to reflect the current capabilities of IT systems, particularly GCSS-MC, TFSMS, and DRRS.

While we conducted a comprehensive evaluation of current provisional unit equipping-related orders, we also recommend a review of policy letters (particularly at the MARFOR and MEF level) and standard operating procedures (SOPs). The goal of the review is to clarify and standardize processes across the MARFORs, MEFs, and provisional units while ensuring that the Marine Corps retains some degree of policy flexibility for unforeseen future circumstances.

Forward-Positioned, Flexible Equipping Practices

One striking fact that we observed is that some provisional units have access to certain equipment pools and others do not.[2] Inconsistencies are often related to different policies governing the use of equipment pools, some of which are forward deployed or considered war reserve

---

[2]   See discussion in Chapter Two regarding SPMAGTF-CC-CR's regular use of the MAP-K in contrast to MRF-E's difficulty in using MCPP-N.

materiel. The Marine Corps should reconsider using the variety of equipment pools that currently exist—for example, Combat Ready Storage Programs (CRSPs), MAP-K, MCPP-N, and MPS—to equip provisional units. Although maintenance and turnover policies will need to be refined or developed, using excess equipment pools to equip provisional units can mitigate the readiness risks associated with using regular unit equipment and thus formalize a practice that is already being followed for some provisional units.

As the Marine Corps goes to more distributed operations, as indicated in the *Marine Corps Operating Concept*, it will need to have more logistics flexibility to operate in austere environments. There are plenty of lessons to be learned from the provisional units that are useful in this operating environment, such as the Marine Corps

- enacting potential broader acquisition authorities to enable units to buy commercial off-the-shelf gear
- using technology such as 3-D printing to supplement the traditional supply systems
- leveraging partnerships, such as what MRF-D is doing in Australia to fix tires and corrosion with Australian partnership organizations
- pulling from prepositioned stocks while limiting risk to full-scale operations.

Improved Communication and Visibility Across IT Systems

To support many of the above-mentioned policy recommendations, the Marine Corps needs to improve GCSS-MC functionality so that it ensures provisional unit equipping visibility, in addition to ensuring data accuracy. Provisional unit EDLs, the equipment owner, and maintenance and modernization needs should be clearly visible to all users, which is not the case today. It also needs to speak more clearly to other enterprise systems such as TFSMS and DRRS-MC. Among other advantages, upgrading current supply and equipment IT systems will ensure that mission needs are accounted for over time and support accountability when using equipment pools for provisional unit equipping.

Elimination of PEI Rotation Policy by Instead Updating Other Marine Corps Policies

Although the PEI rotation policy was conceived to address provisional unit equipping issues, it has not proved to be effective and should be discontinued. The current iteration of the policy reduces equipment availability at provisional units without substantially improving readiness or modernization. We assess the adoption of these policy recommendations will preclude the need to maintain a strict 20 percent rotation requirement. Rather, the recommendations, if adopted, will drive the Marine Corps toward a more needs-based approach to maintenance and modernization.

Hybrid Approach of Courses of Action Evaluated

After evaluation of the four equipping strategies, no single course of action emerged as solving the majority of the equipping challenges identified. Instead of choosing a single strategy for implementation, we recommend that a hybrid approach that capitalizes on the strengths of each of the COAs be pursued.

To begin with, have MARCORLOGCOM provide a centralized management function to manage excess equipment inventory and remove the burden of redistribution from the MEFs. To address the HD/LD shortfalls, something similar to the permanent structure COA would be recommended, but in lieu of increasing the AAO for all provisional equipment, the Marine Corps could selectively procure additional HD/LD items to alleviate the demand. Finally, the Marine Corps already conducts aspects of the tailored equipping COA by placing equipment sets in forward-deployed locations (MAP-K, MCPP-N, MPS). Making it easier to pull from these pillars of the AAO could assist in alleviating some problems faced by provisional units.

Conclusion

In documenting the extent of the equipping issues provisional units face, the research team expected to find a small number of underlying issues that were causing most of the overall problems that were initially identified by HQMC, I&L. However, this was not the case. Equipping problems stemmed from many small problems that collectively caused negative impacts. We realized that no single course of action would significantly improve provisional unit equipping. Instead, the problem demands a multipronged approach to solve the totality of the problems. The result of this realization is the collection of distinct recommendations in the chapter above. Therefore, we recommend the full adoption of all the recommendations presented in this chapter, in the following sequence:

- In the near term, eliminate the PEI rotation policy. Put in place a plan for conducting empirical assessments to determine actual equipment usage. Update policies to reflect current capabilities and address sustainment challenges. Couple this update of policy with improvements to IT systems to better support policies as written.
- Once the Marine Corps has a better understanding of equipment usage and has updated policies and systems, implement the use of risk matrices to prioritize equipping solutions for different provisional units. Implement the hybrid COA in a manner that takes into account the risk matrix, and implement flexible and forward-positioned equipping strategies.

Interview Protocols

Two interview protocols were used during the project, one for personnel within provisional units, another for personnel associated with units supporting provisional units. The supporting unit protocol was also adapted for use with personnel associated with the supporting establishment.

Provisional Unit Protocol

Contextual questions
1. Collect indexing data
   a. Name
   b. Rank
   c. Unit
   d. MOS (military operational specialties)

How does your provisional unit source personnel and equipment requirements?
1. Can you provide us a copy of your equipment density list?
   a. What units source equipment for your EDL? How was it chosen? By whom?
   b. How were your equipment requirements determined?
   c. Were there any issues with receiving the equipment you required for deployment? E.g., maintenance, variants, substitutes.

d. Does your unit have the right equipment to conduct its mission?
e. Does your unit have excess equipment that is not utilized?
f. What equipment did your unit use to train for this deployment? And where?
g. Did you have adequate/appropriate equipment to train with?
h. How do you track the movement of equipment into and out of your unit? How do you maintain visibility of your equipment?
i. What advantages/disadvantages are there in the current method your unit uses to maintain visibility?
j. What equipment challenges have you encountered during your deployment?
2. Can you provide us a copy of your manning document?
a. What units source personnel for your manning requirements? How were they chosen? By whom?
b. How were your manning requirements determined?
c. Were there any issues with receiving the personnel you required for deployment?

What is the impact on home station units?
1. What impact do sourcing requirements for your provisional unit have on the ability for home station units to conduct their mission?
a. Impact on readiness to meet operation plan and contingencies?
b. Impact of supply and maintenance readiness?
c. Impact on reset efforts?
d. How do you know/track these impacts?
2. How has your unit been coping with that impact?
a. Are those measures effective?
b. What measures have you considered and discarded?
c. What measures have you considered but not yet tried?

What can be done to solve the problem?
1. If you were given a clean sheet and tasked to develop a support mechanism for provisional units that balances the needs of BOTH units, how would you do it?

2. If you could change *one thing* about the equipping of SPMAGTFs, what would that be?
3. Any other thoughts or issues that we have not covered?
4. Are there any other individuals or groups that we should contact?

Supporting Unit Protocol

Contextual questions
1. Collect indexing data
   a. Name
   b. Rank
   c. Unit
   d. MOS
2. What provisional units do you support?
3. How long have you been supporting the provisional unit(s)?

What is the scope of the problem?
1. How does the MEF support provisional units?
   a. Does your support change by unit and/or mission?
   b. Please walk us through the entire cycle of support you provide both for equipment and personnel.
2. How do you maintain visibility over the support you provide?
   c. EDLs?
   d. Other methods?
   e. What disadvantages are there in the current method your unit uses to maintain visibility?
   f. Does the current process for accountability differ from what is laid out in policy/doctrine? If so, why?

What is the impact of the problem?
1. What impact does supporting provisional units have on MEF readiness?
   a. Impact on readiness to meet operation plan and contingencies?
   b. Impact of supply and maintenance readiness?
   c. Impact on reset efforts?
2. How has the MEF been addressing these readiness issues?

    d. Are those measures effective?

    e. What measures have you considered and discarded?

    f. What measures have you considered but not yet tried?

What can be done to solve the problem?

1. If you were given a clean sheet and tasked to develop a support mechanism for provisional units that balances the needs of BOTH units, how would you do it?

2. Any other thoughts or issues that we have not covered?

3. Are there any other individuals or groups that we should contact?

Qualitative Methodology

RAND used a qualitative research method known as thematic analysis to gain deep understanding of the provisional unit equipping problem. In thematic analysis, researchers review qualitative data (such as interview transcripts) for recurring patterns across interviewees and perspectives.[1] These patterns, or themes, are first identified and articulated by the research team before approaching provisional unit equipping stakeholders and refined throughout the interview process. Thus, we can articulate hypotheses about provisional unit equipping problems and confirm, refute, or refine them iteratively through the interview process. Additionally, we constructed several open-ended themes (i.e., any comment that mentioned a type of provisional unit) that allowed us to identify trends without having to articulate them before analysis. Thus, we combined inductive and deductive approaches to comprehensively analyze the provisional unit equipping problem.

We interviewed 24 provisional unit equipping stakeholders for this research. These stakeholders represented a range of viewpoints, from provisional unit logistics officers, to higher-echelon logistics staffs responsible for managing equipping, to HQMC organizations that write or influence equipping policy. A list of the stakeholder organizations interviewed is included in Table B.1.

---

[1] Although substantial quantitative data were available for analysis, we understood that many provisional equipping problems were difficult to separate from the context of the organizational environment that these problems occurred in; this is an ideal situation for using thematic analysis.

Table B.1
Stakeholder Organizations Interviewed

Provisional Units	Headquarters Marine Corps	Operating Force	Supporting Establishment
SPMAGTF-CR-AF SPMAGTF-CR-CC MRF-D MRF-E	HQMC, I&L HQMC, PP&O HQMC, CD&I HQMC, TFSD	I MEF III MEF 7th MARFORPAC 4th Tank Battalion (Reserve Unit) Combat Logistics Regiment-1 MARCENT MARFORCOM	MARCORLOGCOM ESD

SOURCE: RAND.

NOTE: Interviewees were determined in coordination with HQMC, I&L and the Marine Corps Operations Analysis Directorate.

Interviewee Demographics

Each interview was assigned certain descriptors to understand the demographic makeup of those interviewed. Those demographic descriptors are illustrated in Figures B.1, B.2, and B.3.

Figure B.1 demonstrates that interviewees came from across the Marine Corps enterprise, but the majority of interviews focused on the input from provisional units themselves. This was followed closely by HQMC personnel. Figure B.2 breaks the list of interviewees down by billet type. Here there was a focus on logistics personnel who were responsible for provisional unit equipping and interacted the most with the process. Their input was augmented by feedback from commanders and operations personnel. Finally, Figure B.3 shows the seniority of personnel interviewed.

Figure B.1
Interviewees by Organizational Role

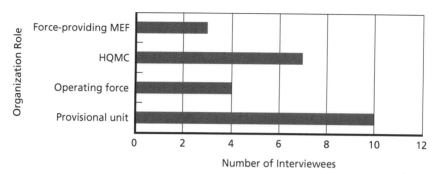

SOURCE: Dedoose analysis of SME interviews.

Figure B.2
Interviewees by Billet Type

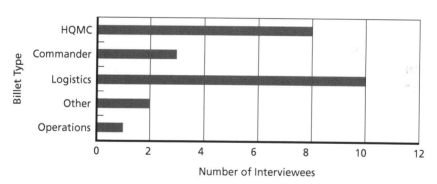

SOURCE: Dedoose analysis of SME interviews.

Figure B.3
Interviewees by Rank

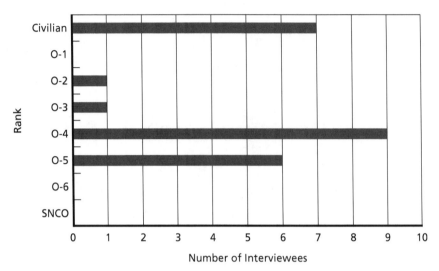

SOURCE: Dedoose analysis of SME interviews.

Themes Identified for Analysis

As mentioned above, we articulated two types of themes for this analysis: themes related to the initial hypotheses that we had about the provisional unit equipping problem and open-ended themes that allowed us to explore other possible hypotheses and theories not previously known to us. Table B.2 lists the themes that we identified, with hypothesis-driven themes marked with an asterisk.

Table B.2
Initial Hypotheses Themes

Theme Category	Theme	Definition
Current equipment practices	CRSP	Any mention of the Combat Readiness Storage Program
	EAP	Any mention of the Equipment Allowance Pool at 29 Palms, CA
	MAP-K	Any mention of the MEU Augmentation Pool-Kuwait
	MCPP-N	Any mention of the Marine Corps Prepositioning Program-Norway
	Other	Other practices of deploying and rotating equipment not already listed (e.g., UDPs and MEU equipment rotation)
	PEI Rotation	Any mention of PEI rotation
Follow-on effects to provisional unit equipping	Cost*	Any discussion of the cost of the current processes and procedures in place for provisional unit equipping
	Home unit readiness*	Readiness impacts to the sourcing unit as well as the Marine Corps in general
	Lack of efficiency and effectiveness*	Any discussion of problems that cause the equipping system/process to be inefficient or ineffective.
	Operational risk*	Any discussion of loss owing to failed systems

Table B.2—Continued

Theme Category	Theme	Definition
	Provisional unit readiness*	Any discussion of the readiness and availability of deployed unit
	Schedule*	Any discussion of the amount of time various processes take
	Commander's intent*	Any mention of the decisions made at the discretion of a commander
	Culture*	Any cultural barriers that determine what and how things are done
	Equipment sourcing*	Any discussion of current sourcing of provisional unit equipment sets
	Equipment visibility*	Any discussion of the ability to track equipment (i.e., where it is, its condition, where it is assigned)
	External constraints*	Anything that impacts the provisional units that are outside of I&L's control
Provisional unit equipping problems	Information systems	The systems and tools used to track and process units
	Maintenance*	Anything related to maintenance capability (e.g., availability, maintainers, spare parts)
	Mission*	Mission requirements that impact equipment use and availability
	Operations*	Anything related to the ability to conduct operations
	Personnel*	Any discussion of the number of personnel or skills of personnel
	Policies and procedures*	Any mention of the policies and procedures that govern provisional unit equipping
	Training*	How training is impacted because of certain equipment availability
	Transportation*	When, why, and how equipment is moved between locations/units

Table B.2—Continued

Theme Category	Theme	Definition
Provisional unit	BSRF	Specific concerns related to this unit
	MRF-D	Specific concerns related to this unit
	MRF-E	Specific concerns related to this unit
	Other unit	Specific concerns related to this unit
	SPMAGTF-CR-AF	Specific concerns related to this unit
	SPMAGTF-CR-CC	Specific concerns related to this unit
Reactions to COAs	Against	Disagree that a specific COA would work, along with reasons why
	For	Agree that a specific COA is a good idea and could work in the Marine Corps
	Neutral	Interviewee undecided and/or sees pros and cons to a specific COA
	Strengths	Regardless of interviewee's overall opinion of the COAs, this code identifies potential strengths of a certain course of action
	Weaknesses	Regardless of interviewee's overall opinion of the COAs, this code identifies potential weaknesses of a certain course of action
Recommendations and COAs	Add to AAO	Any discussion of adding provisional unit equipment to the AAO
	Other recommendation	Recommendation an interviewee suggests that does not fall into one of the COAs
	Centralized management	Any discussion of centrally managing provisional unit equipment
	Tailored equipping	Any discussion of tailored equipping
	Sufficient	Anything involved with provisional units that is considered sufficient/ working well enough (applied on top of child codes associated with "problems" if the problem area is indeed not a problem)

SOURCE: Dedoose analysis of SME interviews.

NOTE: EAP: equipment allowance pool; UDP: unit deployment plan.

*Hypothesis-driven theme.

Analysis Conducted Using Dedoose

RAND used Dedoose, a thematic analysis software tool, to analyze the data set. Dedoose allowed us to systematically link interview transcript comments and responses to themes (a practice known as "coding") and relate them to one another. For instance, Dedoose allowed us to examine a theme—operational risk, for instance—and examine all mentions of this theme across stakeholder perspectives. This allowed us to identify themes that resonated most strongly with different groups of interviewees, giving us a powerful tool to confirm or deny hypotheses.

As Dedoose is a cloud-based application that allows for collaborative coding, we employed a number of trained coders to analyze the collected interview transcripts. Coders were varied in familiarity with the provisional unit equipping but were uniformly trained and tested for between-coder agreement before being randomly assigned interview transcripts to code.

Cost Analysis

Earlier we evaluated each course of action for strengths and weaknesses based on subject matter input and team analysis. Additionally, we worked with key stakeholders to develop cost estimates with enough fidelity to adequately ensure discrimination between courses of action. The deliverables as a result of the cost analysis will include a rough order of magnitude for annual costs to the Marine Corps, including any personnel, facilities, and maintenance costs, for each of the alternatives. The Government Accountability Office (GAO) Cost Estimating and Assessment Guide was used as a resource to ensure a standardized approach to the development of the cost estimates and associated analysis. Ultimately, following the initial analysis, the selected COA will require a more robust cost estimate.

Cost Discussion Ground Rules and Assumptions

We developed a set of preliminary ground rules and assumptions for the cost analysis. The initial set of ground rules and assumptions was global in nature, capturing items such as constant year versus then year dollars, inflation indices to utilize, fiscal years to include in estimate, non-materiel solution descriptions, and the bounded scope of the technical solutions. As alternatives were developed, more alternative and specific ground rules and assumptions were added. The ground rules also established a clearly defined standard cost element structure, based on the areas where there will be the most variation in costs among alternatives, and will clearly lay out the set of costs to be included or not included as part of the COA assessment process. Among the assumptions:

- The estimates will not consider any legacy or sunk costs.
- The EDLs for provisional units are relatively constant over time.[1]
- There is natural variation in costs year to year based on changing strategy, operational tempo, and training requirements that will not be captured in these point estimates.
- Time-based maintenance, for exposure to corrosion, sand, or other resets, does not get prioritized by provisional units today and therefore happens less frequently than for CONUS-based units.
- TAMCNs on the AAO can be procured at I&L standard unit price.[2] Therefore, these estimates do not account for the costs of restarting production lines or have obsolescence challenges.
- Ratio of weight/cube to value is similar for Crisis Response-Central Command (CR-CC) as other provisional units and therefore can be used to predict costs of overseas shipments.
- Rates for cube provided by USTRANSCOM would on average be representative of shipping costs to and from provisional units and sourcing units.
- Personnel costs for MAGTF equipment allowance pool (EAP)[3] are on average per full-time equivalent (FTE), similar to tailored equipping.
- Provided maintenance data are comprehensive.

---

[1]   HQMC, "Task-Organized Unit EDLs.xlsx," October 3, 2017.

[2]   HQMC, "Force 2025 AAO vs Inventory UIC Level-Data Model.xlsm," October 19, 2017. We used the I&L data to derive an average standard unit price for each item. Each TAMCN may have more than one associated NIIN, and each NIIN may have a different standard unit price. In order to conduct consistent analysis at the TAMCN level, we developed TAMCN standard unit price based on the average of associated NIINs, weighted by the quantity of those NIINs currently on hand.

[3]   United States Marine Corps Operations Analysis Directorate Combat Development and Integration, "MAGTF Training Command Equipment Allowance Pool Study," Quantico, Va., April 5, 2017.

Cost Discussion Methodology

The estimating methods selected vary depending on a variety of factors, such as the level of confidence of the input parameter values, projected schedules, level of detailed data available, and currency of the information captured. The cost estimating method chosen largely depends on the maturity of the solution and data availability. Currently, the courses of action are notional and existing data sets have known deficiencies. Thus, the personnel, operations and maintenance (O&M), and disposal cost estimates for each alternative are based on a tailored mix of engineering actual costs and analogy-based estimating techniques. Where necessary, expert opinions were also solicited. All collected data were normalized in the most appropriate manner, including adjusting for inflation, quantity, time frame, and customizing for technical characteristics where appropriate across each of the alternatives.

Cost Discussion Data and Tools

The team first solicited data on existing status quo process costs to include both materiel and non-materiel costs. As alternatives were further developed, our team requested maintenance data from Visibility and Management of Operating and Support Costs (VAMOSC) and Total Life Cycle Management-Operational Support Tool (TLCM-OST),[4] budget data for provisional units, transportation cost data, and number of personnel. Data collection also consisted of reviewing other related studies for creating training sets and changing provisional unit materiel sourcing. We used discussions with personnel familiar with provisional unit operations, interviews with SMEs, and research of open-source information to round out the data.

The cost model was built as a tailored mix of estimating techniques for quantifying the differences in complexities across the alternatives, but as many Marine Corps data sources have known limitations, our analysis is a very rough order of magnitude estimate of personnel, transportation, maintenance, acquisition, and facilities costs. With

---

[4]   United States Marine Corps, Total Support Cost Dashboard, 2018.

these parameters in mind, the next section presents the analysis for each of the COAs.

Centralized Management

Personnel Costs
There will be an increase of personnel needed at MARCORLOGCOM to centrally manage this equipment. This would be best calculated by adjusting the staff from a similar MARCORLOGCOM-managed program. During this effort we were unable to identify an analogous program and the appropriate staff mix, but we anticipate that it would be less staff than estimated for the tailored equipping strategy due to the ability to leverage existing manpower resources already present within LOGCOM as opposed to creating new requirements.

Transportation Costs
For centralized management, there will continue to be requirements to move materiel to and from CONUS to ensure provisional units have the right mix of equipment and that equipment is ready. In this course of action, we assume less transportation than estimated for conducting the PEI rotation where one-fifth of the materiel is being moved to and from CONUS each year. Thus, we estimate it will be under $8 million per year, but there may be some initial increase in movement as the centralized management works to ensure equipment is ready and available to provisional units.

Maintenance Costs
Under a centralized management regime, usage-based maintenance will likely remain constant.

Acquisition Costs
There will be no immediate changes to equipment on EDLs to support this course of action. Therefore, there will be no additional cost associated with acquisition.

Facilities Costs

There will be no changes to the number or mix of facilities to support this course of action. Therefore, there will be no additional cost associated with facilities.

Permanent Structure

Personnel Costs

While permanent force structure would also be recommended in conjunction with permanent T/E, the personnel aspect of provisional units was outside the scope of this project. We estimated costs for personnel in terms of support personnel required for the equipping portion of this COA. There will be no changes to the number or mix of personnel to support this course of action. Even though there will be increased maintenance requirements because of an increased amount of equipment, without a permanent change in T/E there would be no increase in uniformed personnel. Therefore, there will be no additional cost associated with personnel unless permanent force structure is authorized.

Transportation Costs

In this COA there will be some initial movement of equipment to provisional units in the millions of dollars. In addition, there will be a correlated ongoing increase in annual transport requirements. To estimate this cost, it is critical to understand the cube information for the equipment that is added and the origin and destination locations for all materiel. Once those data are collected, USTRANSCOM will provide an estimated cost for the movements.

Maintenance Costs

When the amount of equipment increases, it is likely that the usage for the equipment will also increase because of availability. Also, there will be corresponding costs for time-based maintenance (e.g., for corrosion and exposure to sand) that will need to be handled for the additional pieces of equipment in provisional units.

Acquisition Costs

The cost to increase the equipment to the AAO depends on whether the USMC can reuse excess already available. If the USMC recaptures this excess, the cost will be approximately $200 million using the I&L price for the TAMCNs that would need to be purchased. If all TAMCNs are purchased new rather than being obtained from excess the cost is about $450 million. Similar costs were calculated by a previous study.[5]

Facilities Costs

There will be no changes to the number or mix of facilities to support this course of action. Therefore, there will be no additional cost associated with facilities.

Tailored Equipping

Personnel Costs

To support tailored equipping there will be an extensive equipment pool that requires logistics support. Similar to the ESD of the EAP at 29 Palms, CA, civilian support staff would be required to manage the equipment pools. Given this requirement, we took the initial estimates for a similar equipment pool[6] and adjusted them to reflect the smaller size and different assets required for this mission. The estimated civilian personnel and their associated annual costs for 2018 totaled approximately $10 million. The amount of military personnel was approximately $5 million.

Transportation Costs

The cost of transportation is dependent on the original location of the items being added to the tailored equipping set. Once those item loca-

---

[5] Department of the Navy, Marine Corps Requirements Oversight Council, "MROC Decision Memorandum," February 2016.

[6] United States Marine Corps Operations Analysis Directorate Combat Development & Integration, "MAGTF Training Command Equipment Allowance Pool Study," briefing, Quantico, Va., April 5, 2017, pp. 9–12.

tions, weight, and cube are identified, analysts can obtain quotes for movement.

Maintenance Costs

To calculate the impact on maintenance, the estimator needs data on operational tempo. Once those data are available it is possible to estimate the cost of usage increases using data from TLCM-OST or unit data to understand the increases in maintenance owing to usage and time-based maintenance requirements.

Acquisition Costs

There will be no immediate equipment acquisitions to support this course of action. Therefore, there will be no additional cost associated with acquisition. However, there should be long-term savings associated with a smaller overall Marine Corps equipment requirement.

Facilities Costs

The facilities cost estimate is an analogy to a similar equipment set estimated for installation at Yuma, Arizona. For this estimate, we took the MAGTF EAP's[7] estimate and adjusted the costs using the most recent Unified Facilities Criteria (UFC) cost factors and uncertainty.[8] Then we estimated the facilities to be about 75 percent of the size of the proposed Yuma facilities. The UFC model was run with the original square footage and multiplied by the 75 percent factor. Then, to bound the uncertainty of the results, the UFC model was rerun with reduced square footage. These two methods resulted in an estimate for $28 million investment, plus or minus $6 million in uncertainty, and an annual upkeep of $400,000, plus or minus $75,000 in uncertainty.

---

[7]   United States Marine Corps Operations Analysis Directorate Combat Development and Integration, "MAGTF Training Command Equipment Allowance Pool Study," April 5, 2017, pp. 7–8.

[8]   Department of Defense, "DoD Facilities Pricing Guide with Change 1," May 23, 2018.

Summary

It also should be noted that the cost analysis was limited by the availability of data. We recommend that the Marine Corps work to capture data that would allow for more detailed analysis. Primarily the service would benefit from more robust information on operational tempo and equipment usage in order to improve cost estimation. Better information on disposal and transportation costs would also be useful.

References

Clinger, Richard, "TAMCN Dimension Data," email to Joslyn Fleming, Quantico, Va., June 5, 2018.

Department of Defense, "DoD Facilities Pricing Guide with Change 1," May 23, 2018.

Department of the Navy, Marine Corps Requirements Oversight Council, "MROC Decision Memorandum," February 2016.

Harkins, Andrew, "SPMAGTF 15-Aug-2017 to 4 -Jan-2018 Source of Supply Stats.xls," email to authors, March 9, 2018.

Headquarters Marine Corps, *Marine Corps Planning Process*, Marine Corps Warfighting Publication 5-1, Washington, D.C., August 24, 2010.

———, *Marine Corps GFM and Force Synchronization Manual*, Marine Corps Order 3120.12, Washington, D.C., February 11, 2015.

———, *Total Force Structure Process*, Marine Corps Order 5311.1E, Washington, D.C., November 18, 2015.

———, *Management of Property in the Possession of the Marine Corps*, Marine Corps Order 4400.201 Volume 3, Washington, D.C., June 13, 2016.

———, *Marine Corps Operating Concept*, Washington, D.C., June 2016.

———, "Force 2025 AAO vs Inventory UIC Level-Data Model.xlsm," HQMC, I&L, October 19, 2017.

———, *Marine Corps Readiness Reportable Ground Equipment*, Marine Corps Bulletin 3000, Washington, D.C., March 21, 2017.

———, "Task-Organized Unit EDLs.xlsx," HQMC, I&L, October 3, 2017.

———, "AAR- IPR#2 on RAND Analysis of Provisional Unit Equipment Management," HQMC, I&L, May 9, 2018.

———, "HD/LD Equipment List.pptx," HQMC, I&L, February 7, 2018.

Lewis, Matthew W., Aimee Bower, Mishaw T. Cuyler, Rick Eden, Ronald E. Harper, Kristy Gonzalez Morganti, Adam C. Resnick, Elizabeth D. Steiner, and Rupa S. Valdez, *New Equipping Strategies for Combat Support Hospitals*, Santa Monica, Calif.: RAND Corporation, MG-887-A, 2010. As of November 18, 2018: https://www.rand.org/pubs/monographs/MG887.html

Pernin, Christopher G., Edward Wu, Aaron L. Martin, Greg Midgette, and Brendan See, *Efficiencies from Applying a Rotational Equipping Strategy*, Santa Monica, Calif.: RAND Corporation, MG-1092-A, 2011. As of November 18, 2018: https://www.rand.org/pubs/monographs/MG1092.html

Turner, Brian, Anne-Marie Adams, Dr. Raphael Laufer, and Stephanie Woodring, *Program Objective Memorandum-19 Front End Assessment; Aligning Ground Equipment Inventories to Operational Requirements*, Washington, D.C., September 30, 2016.

United States House of Representatives, *Statement of General Robert B. Neller Commandant of the Marine Corps: Hearing Before the House Appropriations Subcommittee on Defense on the Posture of the United States Marine Corps*, Washington, D.C., March 7, 2018.

United States Marine Corps, Total Support Cost Dashboard, 2018.

United States Marine Corps Centers for Lessons Learned, *Logistics Combat Element in Support of Special Purpose Marine Air Ground Task Force—Crisis Response—Africa 15.2 and 16.1 Rotation; Combat Logistics Battalion 6*, Washington, D.C., September 20, 2016.

United States Marine Corps Operations Analysis Directorate Combat Development and Integration, "MAGTF Training Command Equipment Allowance Pool Study," Quantico, Va., April 5, 2017.

———, "Performance Work Statement for the Analysis of Provisional Unit Equipment Management," Quantico, Va., June 23, 2017.

Vinyard, Bill, "Data Access," email to EGEM and RAND staff, Quantico, Va., October 18, 2017.

Wiesemeyer, Kyle J., "Cost Estimate Request," email to USTRANSCOM, Scott Air Force Base, Il., June 13, 2018.